Dedication to Skilled Tradesmen

This book is dedicated to skilled tradesmen—past, present, and future. Your hard work and craftsmanship have built the world around us and inspires new generations to pursue skilled trades.

I Built That! And So Can You is a must read for anyone interested in a career in the construction industry. It is chock full of real-world experiences on the jobsite and in the fab shop. This book is a real life look at how diversified, challenging and rewarding a job in the construction industry might be. Mr. Hoover's industry experience is a shining example in every chapter of the book.

Phil Petersen, Chapter Manager Inland Northwest Sheet Metal Contractors Association, Spokane, Washington

Roger Hoover's book, ***I Built That! And So Can You*** is an easy, engaging read for anyone. As a fellow tradesman, its stories energize me and reflect experiences I can relate to throughout my career. Recommended reading for parents of prospective building tradesmen.

Brawny Welch, SMART Local #24
Dayton Business Representative

Having been employed in the skilled trades for over 50 years, Roger provides unique insider experience of the opportunities and at times adventurous side of the skilled trades as only he can tell. A timely book written by a 72-year-old, well respected in his community and on the job site. An enjoyable read for anyone considering an option to attending college.

Michael Petro,
Allamericanstore.us

Roger Hoover's book, ***I Built That! And So Can You***, stands as a powerful testament to the importance of skilled trades education and its undeniable impact on the broader economy. It dismantles the outdated myth that skilled trades jobs are low-paying, redundant, or lacking in respect. These are high-quality careers that young people can enter quickly and confidently without taking on college debt.

Hoover highlights how these professionals go on to build facilities, infrastructure, and public works that they can take pride in for a lifetime. He accurately and eloquently outlines the current challenges facing the skilled trades and offers a thoughtful path toward addressing the shortage of these essential workers and their specialized skills.

Dr. Nick Weldy, Superintendent,
Miami Valley Career Technology Center

Thank You

Huge thanks to all the skilled tradesmen I have shared time with on the job. It was not possible to complete this work without your shared time "hammering" out my years in the trade.

I am grateful to my family and friends for patiently listening to countless stories throughout my long and fulfilling career.

To my wife, for her patience and honest feedback as I pursued this dream.

To my four sons and grandchildren who mean everything to me and always asked about my "book."

Deepest appreciation to Jacki Dolan, whose editing allowed this book to come to life and become readable, logical and enjoyable. Her input shaped much of the final content.

I want to express my gratitude to my publisher and his team at Braughler Books. From our first meeting to discuss the project, all the way to when I received my finished book, David Braughler has been fantastic to work with. As a first-time author, I've been impressed by his thorough understanding of the publishing world. It's been a pleasure to experience this journey with him.

A Look At Construction And Skilled Trade Opportunities In The Twenty First Century

I Built That! And So Can You

Roger Hoover

Cover photo of the Generations Statue courtesy of the Sheet Metal Workers' National Pension Fund.

Printed in the United States of America

First Printing, 2025

Softcover ISBN 979-8-89390-071-2
Hardcover ISBN 979-8-89390-101-6

Library of Congress Control Number: pending

Ordering Information: Special discounts are available on quantity purchases by bookstores, corporations, associations, and others. For details, contact the publisher at sales@braughlerbooks.com or at 937-58-BOOKS.

Braughler Books
braughlerbooks.com

For questions or comments about this book, please write to info@braughlerbooks.com.

Contents

Introduction

This book is about the construction industry, an industry that I dearly love and have worked in for over fifty years. It is an industry that has changed dramatically over the years. It is one that is built on tradition and hard work; it has moved from a very dangerous industry to a very safe one. The business has changed from a direct problem-solving contact with the customer to one of no contact with the customer at all. It is one that has seen the control change to a ridiculous level of absurdity. To eat lunch requires the worker to have safety glasses on, hard hat worn and gloves required. It is one that had first responders to a jobsite (for an ambulance call for a possible heart attack) stopped by the construction manager, told that they were not allowed on site without hard hat and safety glasses. Despite all of this, it is still an industry and way of life that I continue to love.

As I see it, there are two types of people in the world: ones who shower *before* they go to work and those that shower *after* they come home from work. People who shower first are those that work as teachers, doctors, attorneys or judges; those that need to have a proper presence in the workplace and a sharp appearance. And there are those that work first and shower later. They are the building trades and service industry. Those that get dirty during working hours: the plumbers, electricians, ditch diggers, power workers and sheet metal workers. They expect to get dirty and dress accordingly. They may work outside and in terrible working conditions. They are also some great people with a fabulous work ethic and can be highly paid for their work. I fell into the "shower-after-work" group for many years. This is my story.

Chapter One

My Story

I am a storyteller. This is a story of construction, education, and everything in between. It is my story as I remember it, so it may not have happened exactly as it occurred.

It all began on June 20th in 1972 when I became a sheet metal apprentice at the Kirk and Blum Manufacturing Company. I applied for an apprenticeship through Local 224, Dayton, Ohio, affiliated with the Sheet Metal Workers International Association. Little did I know where this would take me. I developed an interest in the field of construction through an outreach program of the local union that was presented at my high school in Dayton. It was also a time dedicated to making the trades more accessible to minorities. I took a test to qualify for the opportunity at the local union hall and placed first out of nearly two hundred applicants. I contacted the apprenticeship office to check my results and was surprised that the administrator knew my ranking without even checking. I had achieved the top position on the list and was offered a job immediately, should I choose to accept.

At the time, I was working with my father painting houses. My father was a full-time high school teacher and during the summer worked with other teachers painting houses. Teachers with a four-year undergraduate degree and a master's degree still needed to work for extra income to make ends meet. I enjoyed working with my father, so it was hard to leave so abruptly. But off I went into the unknown.

The first day as an apprentice included a tour of the small office and shop. I was as green and unskilled as you could ever imagine. I had never worked in construction and had no experience in the trade. Being thrown into a world that was quite unfamiliar to me, I was in a learning experience that was like drinking from a fire hose. The tour included meeting the small group of workers in the shop. The shop was driven by the skilled knowledge and experience of workers that hand maneuvered through what is now computer assisted in the sheet metal shop of today. Each of the workers had his own unique story of how he started in the trade. Every day brought new and different tasks to learn and remember. I was just trying to work through a new environment and a new language, "shop talk." I grew up in a family that valued education and was well spoken. My family did not swear like these construction workers did. X@#*! I remained afraid to speak for the next year or so.

By the time I started apprentice school in the fall, my on-the-job learning had put me ahead of my fellow apprentices. My class included some who took extra classes to pass the entrance test and several minorities who fit well in the program. The construction industry does not tolerate much when it comes to offering something for nothing. You work or you are dismissed. Easy enough. No case building, no seniority, no toleration for anything but production.

The year before I started working in construction had been slow and no apprentices were accepted into the program. I recently graduated from high school and decided to attend college to become a medical technologist. Classes (biology, science, chemistry) were demanding, but I was progressing through the system. I really did not want to go to college, so when the apprentice opportunity presented itself again, I moved toward the chance to get into the industry. I enjoy working with my hands and being able to see what you accomplish at day's end.

It was not a hard decision to move into construction the following year. In 1972 I began earning $4.30 an hour, which was fifty percent of a journeyman's salary. Because our contract ended on June 30, I received a raise after only ten days on the job. The union wage when I started was $8.60 an hour. Half of $8.60 was not as good as the $5.00 that my dad paid me to paint. But the future would result in better pay and full benefits. And at the time, the minimum wage was $1.25 an hour.

As the apprenticeship classes in sheet metal moved forward, I continued to improve and gain new skills. Many hours of work on the job prepared me like no other way could. All the journeymen appreciated my willingness to work, and work we did. My first day off was the first weekend in October. I had worked straight through since June and received a full pension credit for my first year as I had the minimum 1600 hours worked in less than six months. That averages out to almost nine hours per day for six months.

My apprenticeship was an awesome way to learn the trade as I was working at a company that highly valued education. Shortly after

I started, the draftsman moved into the shop foreman position, and I was tapped for the drafting job as I had some limited experience in drafting in high school. I was detailing drawings with trade terminology that I could not believe that I was designing. The company had a salesman that doubled as our engineer who really liked working with me. It was a grand experiment for him in working with someone with absolutely no experience. On one of my drawings, I once used a phase "system to be symmetrically identical", which meant that both ends were the same as I had to shorten one side of my drawing due to lack of room on the paper. When the shop foreman reviewed my drawing, he made the comment, "What the hell does that mean?" The same comment resounded from the outside superintendent. "Our people don't know what you are talking about!" I changed it to say: "Both ends are the same."

That job was designed to be for the perpetual motion engine, the Wankel Engine, that General Motors never built. I worked on many very interesting projects.

During the first year of my apprenticeship, I worked on the drawing board for the first eight hours of the day, moved to the shop for the next two to four hours, and then moved to the field on weekends installing many of the same projects. It was a fabulous opportunity as I designed, fabricated, and installed some of the same projects. No blaming the other guy for the mistake. Often, I *was* the other guy.

One project involved designing a forty-inch elbow that angled up to a roof that included a transition in the middle of the run up to the roof. It was made of heavy gauge material and needed to be installed on a Sunday as the duct ran underneath a railroad track and exhausted an industrial oven. The crane had to work around the train schedule as the crane needed to be set on the tracks. The front of the crane was cribbed up about four feet to level out the crane. This involved placing lumber under one end of the crane. Then the ductwork was to be hoisted into position to make the connection. The foreman commented to me that if it did not fit, I was to be fired. No pressure there. After moving some things around, the ductwork fit with the assistance of a big hammer, and I kept my job. Apprentice life can be difficult.

I worked at all seven of the General Motors plants in the Dayton area as we benefited from the many divisions of the auto industry throughout the greater Dayton area. We also specialized in foundry work, and the Dayton area had almost 100 of them. Foundries melt metal and pour it into molds to form many useful metal products such as engine blocks, truck wheels, plumbing fittings and aircraft engine blocks. Kirk and Blum specialized in dust and fume control work associated with the foundry work. They were typically quite busy most of the year. In the 1970s the construction field was still mostly seasonal work. Buildings were not built nearly as fast as today, and crews did not yet have the ability to pour concrete in the winter. Most new construction projects simply waited until the weather improved to move forward, unlike today where the industry is full time with no breaks for winter weather.

In the 1970s, Dayton was booming with a great deal of work. In the early to late seventies, the printing industry was still very strong. McCall's was the largest printing facility in the world under one roof and most major publications were printed in Dayton including *McCall's, Readers Digest, Time, Life,* and *Sports Illustrated*. OSHA (Occupational Safety and Health Act) made its debut in 1976 and generated a great deal of work in many industries as worker safety became a top priority. This generated work in the sheet metal industry as guard work on sanders, polishers, and revolving tools that made any similar work dangerous. The printing industry was a very noisy operation that needed to be made quieter, so the sheet metal industry made the sound panels; the foundry business needed a great deal of work to clean up the dangerous and dirty atmosphere for workers. The sheet metal industry fabricated and installed hoods and dust collectors to solve this problem. It was a good time to be in the service business to ensure safety and to clean up the workplace.

It was also a booming time in the auto industry with seven divisions of General Motors in Dayton. Dayton products made by GM included refrigerators, washing machines, brakes, rubber moldings, air conditioners, shock absorbers, steering wheels, seat cushions and much more. Just south of Dayton, Armco Steel was booming in Middletown, Ohio. Project 600 was a 600-million-dollar job that grew to 900-million dollars and improved the operation of the steel mill to include many new and updated ways of making steel. Project 600 had just been completed and flooded the construction industry with many unqualified and untrained workers. It would take many years to cycle through the workforce as many of these workers entered the trades by merely buying their union cards with no training included. It did not offer an apprenticeship to learn the craft. That project was newly finished when I entered the construction industry.

My training continued on a seasonally busy schedule. We fluctuated from summer labor needs of 75 men to a winter labor need of 10 men. In the winter of 1972, I married, and we quickly started a family. Work, school, and family made for a rush of activities. Apprentice school was during the workday as I was schooled one day every other week during the traditional school year of September to May. We were even paid to go to school. The school was held in the basement of the union hall and was taught by a Navy veteran who tolerated very little from the apprentices. Rules were rules and you followed them without question.

Work duties continued as I was assigned the responsibility to measure and draw up the work to be turned over to the shop to be fabricated. All fabrication was completed by hand at the time without the use of any computer-assisted work to make life simpler. My responsibilities continued to grow even as an apprentice.

The responsibilities included designing dust control systems used in all different types of manufacturing, particularly those in the foundry business. Dayton had many foundries that made products that included wheels formed of steel, aircraft engines cast from alloy steel, large and small titanium aircraft parts, and piping and fittings cast from steel and used in the construction industry.

An experience occurred at one of the foundries that cast many parts and pieces, such as valves and fittings. Many were quite small, but some also were quite large. The larger ones generated dust that needed to be controlled. An engineer from the foundry called and wanted some assistance in designing a system. I worked with him on the phone and talked him through to a solution. Both he and I agreed that the solution should work. I then made a trip to the site to explore the jobsite conditions. When I approached the engineering office to contact the engineer, I was stopped in my tracks with the question: "What are you doing here?" The engineer had expected an older, more seasoned designer. We did not look at the jobsite conditions, because I was asked to leave. Upon arrival back at the office, I was asked how the meeting had gone. I reported what had occurred and that I thought the project was dead.

Our wonderful division manager told me he would handle the customer. Nothing more was said to me. I continued with other projects. In moving forward, I eventually re-contacted the foundry engineer with additional questions, which we worked on by telephone. I then visited the site and eventually solved the problem using our original solution and numbers. Somehow my mentor, and very knowledgeable manager, had worked out a way to involve me in the project, and I gained a valuable friendship for many years with the engineer that was sure I was too young to assist.

That experience was repeated more than once. At the time, I was leading a team of workers who were all older than I, working with managers that were all older than I, and working for customers that were older than I. This challenged me on a daily basis.

I had taken a plant tour when I was a first-year apprentice, that fascinated me. Our company was rather slow at the time and the journeyman that I was assigned to wanted to show me the printing industry. Off we went to visit the offices of McCall's. He grabbed a set of blueprints from the gang box for a project at the nearby General Motors plant. We also grabbed a hard hat, which we seldom used on the site in the plant. We marched into the first office on the second floor and promptly opened the GM prints (used only as a decoy) and started looking at the ceiling. As the occupant of the office surprisingly looked at us, he asked what we were doing in his office. My tour guide and journeyman told him we were looking at some work for the near future. Then my tour guide asked the key question: "What do you do here?"

All people are anxious to talk about themselves. The gentleman bought the paper for the plant, so we learned a great deal about magazine quality paper. We then repeated our tour to the next office and replayed the events. We learned a great deal about ink and colors used to print a magazine. Again, a very willing worker that enjoyed his responsibilities at the company.

We finished off our tour with a trip through the process of taking a picture at the ball game (the Reds were in the World Series in Boston that year), sending the picture to Dayton, editing the picture, sending it to New York to be made into a printing plate, sending it back to Dayton to be installed on the printing press to print the magazine. This process started at a Saturday afternoon game and made it to the press by Sunday morning. We were working on the press on Sunday morning to witness the final stage of many steps that it took to produce a magazine photo. Today electronics and technology make this process much easier and certainly faster.

Because the Dayton area was full of a variety of industries, I was able to move about from industry to industry in a fascinating business that assisted these companies with making a wide variety of products.

I also had much to learn as I was certainly inexperienced. On a nice day in my early education of the industry, I had the job to place many lengths of angle iron onto the rack of one of the company trucks so that the field superintendent could make a delivery to one of our jobsites. I placed the angle on the truck and used wire to tie down the load in place.

The superintendent made it almost to the foundry that he was delivering the load to when he stopped at a traffic light and the load came loose and crashed onto the hood of the truck. At this point, he had to reload the truck in the middle of the street and then continue to make his way to the delivery.

When he returned to the shop, the superintendent was not happy, and the first order of business was to teach me how to load and secure the load on the truck. It was then that I was taught how to properly tie knots that held the load securely to the truck.

As I moved through my new career, I watched steel poured, cars assembled, car parts made, aircraft parts built, fiberglass spun, and veneer cut for plywood. GM made steering wheels, foam cushions, rubber moldings (that go around the vehicle doors), brake parts, air conditioners, and windshield wipers. They assembled trucks and much more. I helped with dust control on grinding airplane props, fume control from the pouring of steel molds for ocean-going ships, and fume control of "monkey" dust used in the purification of acetylene gases. All this for an apprentice that was learning an industry of service to the customer. Additionally, I designed a paper trim system for McCall's that trimmed excess paper from the pages of the magazines, both on the short and long sides of the magazine. All this scrap from the printing presses was gathered and recycled by the train carload.

Along with all the new activities that I was experiencing, I had the fortune to work with many skilled and talented workers. It was also a very different time on the job. Expectations were different for the normal workday. One worker was noticing planes flying overhead on a job where we were working outside to set steel for a large unit. He had asked if I knew where the planes were heading. Of course I had no idea.

The next day he did not show up for work. He did not show up for the entire next week. He finally came back to work and indicated to the foreman that he went to get new boots. He *was,* in fact, wearing new boots. He informed me that he did indeed go to get new boots.... in Florida! Where his girlfriend lived. He also said that I may not have known where the planes were going the previous week, but he did know that the one he was on went to Florida through Atlanta. Workers were different then.

My history of involvement in construction started with a company that had direct contact with the customer. As contractors, we looked at the problem that the customer had presented to us, and then suggested solutions based on our knowledge and experience. Our reputation brought us into direct contact with our potential customers. No outside sources of information were brought in to interfere with our recommendations. Sometimes the customer might check with another contractor to get a second opinion or offer a different way to address the problem.

Many problems were tackled merely based on the experience the contractor had acquired. Problem solving was usually based on the plant engineer's acceptance of the solution offered by the contractor. When we worked at any of the auto plants, we worked directly for the plant engineer. We had developed an outstanding reputation with plant engineers.

Jobs done at a plant with a government engineer are very similar. We would still suggest and have accepted a game plan laid out by the contractor. One such job was for the government in Lima, Ohio, at the Army tank plant. Chrysler Corporation built the A1M1 Abrahms tank at the plant. The job was conducted to all the government specifications that accompanied a job that was performed by the government. The work was completed to all specifications, and I thought that we were done.

The government engineer then asked for a test and balance of the project, which meant that I needed to show the engineer all the performance standards by measuring the air flow and all the values. I performed the test and balance and waited one more month to receive our final payment. Then the engineer decided that I needed to plant grass around the base of our outside unit. That was completed and I waited an additional month. Then he decided I needed to run more tests on the system to create a still more stringent set of values.

I found out that this was the only project that this engineer was working on. He had strung the job out for three more months without final payment. Finally, I approached the purchasing department for our final payment. When the engineering group was made aware of the situation, it was determined that the engineer was milking this lone project for way too long. I returned to his office to inquire about our final payment, only to discover that he was just a *contract* engineer bilking us out of our payment. He had been discharged, and I arrived to find an empty desk. We received our final payment rapidly once the situation was uncovered. I felt terrible that I had not caught on sooner.

Many other projects used an architect or design building engineer. These were primarily duct work jobs. I worked on many of these projects, which are designed by architects and engineers but not completely detailed to include duct lengths or construction. This is up to the individual contractor to complete these details based on *their* production standards if they meet industry standards which are set by SMACNA (Sheet Metal and Air Conditioning Contractors National Association).

On the Cincinnatian Hotel job, I worked as the project manager. The project manager coordinated with other trades to complete the project. The job was an interesting project in that it was to renovate an 1880s style hotel. We were tasked with converting it into a modern and stylish new building. The original building was heated by individual fireplaces located in each room. Each fireplace was vented out through the wall into an exhaust flue that ran through the wall up to the roof. As the building was multistory, the flue became almost as wide as the room when it reached the top floors. The flue was nearly six feet wide at the top. This caused quite some concern as the walls had to support the upper floors.

The building was also very old and needed a great deal of work to support the upper floors. When the architects discovered that the walls were nearly hollow and could not hold any significant weight, it became apparent that new approaches needed to be taken to shore up the walls to carry the additional weight.

My job was to facilitate moving the job forward and included working on change orders. Change orders are required any time the job details change and the need to alter the original contract is apparent. This also includes a need to change the dollar values for the contract.

This style of work removes the customer by one additional layer of management. As a point of contact for the contractor, you now must work through the engineer to get to the general contractor to get to the owner. As you can see, we were moving further away from the decision maker on the job.

I had the opportunity to work with many interesting people over the years. One shop worker was directed to paint a small project in the shop. This should have taken maybe a pint of paint at most to cover the angle iron frame with silver paint.

This journeyman told me, as the apprentice, to watch what was about to happen. He had said to the shop foreman that he did not want to paint this item. The shop foreman instructed him to paint the frame anyway, so he painted the frame, the floor, and the walls, and slopped paint over anything that was within ten feet of the frame. He used about two gallons of paint. Everything within several feet of the angle iron frame was covered with silver paint.

The shop foreman next made the comment, "You will never paint anything in my shop again." The worker simply turned to me with a smile and said, "I told you I didn't want to paint anything." He never painted anything again.

Another funny story that involved an apprentice was on a project to remodel a doctor's office. A junior apprentice was working on the four-story doctor's office building project. He was instructed to go to the roof of the building and take a welding cable with him to hook up a welding machine. He was instructed by his foreman to go to the roof and "throw the cable off the roof."

I was standing next to the foreman as the worker went to the roof, carrying the very heavy cable up four flights of stairs to the roof. Next, he promptly threw the cable off the roof. The cable landed next to the foreman with me at his side. The foreman was furious.

I asked the foreman if I could please ask the worker what he was thinking. As the worker came down the steps onto the ground, I asked him what was going through his mind as he was throwing the cable off the roof.

His response was interesting as he shared his thought process, going through the gestures and acting out his way up the steps repeating over and over, "Why is he asking me to go to the roof and throw the cable off of the roof?"

The foreman only wanted him to throw one end of the cable off the roof but was not specific in his directions. I then asked the foreman what he had told the apprentice to do. The foreman then realized his error in communicating what he wanted to happen.

We often don't mean what we say and *assume* that the receiver of our messages understands what we are saying when they clearly do not. We instruct our crewmembers by verbal instructions nearly one hundred percent of the time even when we know that verbal communication is the worst form of communication of all time. Remember the classic example of sharing a statement through a group of people sitting in a circle, each whispering the message in the ear of the person next to them. The statement changes every time the information is passed along to a new listener.

Very few occasions allow us to put things in written form on the job. As a result, we often miscommunicate with the crew that we oversee in search of a process to clearly communicate as we move toward our goal of completing the daily tasks. Traveling the country and teaching this process also allowed me to gather many examples of miscommunication.

One example was the instruction for a worker to go back after installing some ductwork and complete the installation by turning a one-and-a-half-inch foot on the bottom of the hanger and finish it off by installing a screw in the bottom of the hanger.

The foreman reminded the crew: "Don't forget to turn a foot

on the bottom of the hanger", which resulted in the crew mistakenly turning a twelve-inch bottom to the hanger. A foot. The foreman returned to see a twelve-inch bottom on the hanger with four screws in place. Damn.

Another foreman instructed the crew to go to the roof and drop a tape measure down to the bottom of an existing stack to measure the distance to the bottom. After the crewmember went to the roof, he dropped the entire tape measure down through the stack with the tape measure crashing to the floor. Damn.

I have a son who is a corporate pilot. Early in his career, he was a flight instructor for several years. He received flight instructor magazines at our home for a time, and I read through the material. As the flight instructors were working with students, it was no surprise that the biggest challenge for a flight instructor was miscommunication with a student. The very same problem as a sheet metal apprentice trying to understand his journeyman.

Even though the recommendation was to never attempt to train a family member, I attempted to be trained as a pilot by my flight instructor son. This was the kid who I trained to drive a car, now training me to fly a plane. Unusual dynamics.

One of the challenges of flying has to do with altitude. You really need to stay above the ground to safely fly a plane. There are several different altitudes: above sea level altitude, indicated altitude, pressure altitude, and above ground level altitude. The most important is obviously above ground level.

During investigations into flight accidents, a common cause of accidents is CFIT, *Controlled Flight into Terrain*, which means the pilot flies the plane into the ground. The usual results are not good.

Miscommunication on altitude is very serious.

Miscommunication on the job site is *serious*, too.

One misunderstanding that occurred early in my career was when I just *had enough* of the constant harassment of the journeymen that I worked with every day. It happened while I was working in the shop around many long-time workers that had worked at the company for many years.

On this day, I was certain that I could no longer take the abuse and was putting my tools in my tool bucket to leave and go home and not come back. As I was starting out the door, one of the journeymen stopped me and asked what was happening. I told him I intended to do anything but continue to work at this place.

I would pump gas for a living if I needed to. It was then that the journeyman placed himself between me and the door to the parking lot. He would not move. And he was big. He sternly told me the reason the journeymen at the company were giving me such a difficult time was they all thought I was worth the effort to make my life difficult so that I would appreciate the training they were giving me.

They were, in retrospect, giving me a fantastic opportunity to learn as I was working. They were teaching me and giving me more and better opportunities than the other apprentices. It was hard to see at the time, but I really could not figure out why I got better opportunities than the others. That is a day I will never forget as I finally saw what was happening to me.

Chapter Two

The History of Construction

The history of construction starts with the original building of structures that house everyone at home and at work. Original buildings could not be built over several stories as the walls needed to be thick enough to support the upper floors. Buildings were not built over four or five stories high. This all changed at the turn of the century with the advent of structural steel that allowed structures to be built many stories high. Structures also saw the beginning of organized construction building trades and the original agreements that allowed all involved to agree who would build each component of the structure. Each building trade developed the skills necessary to construct their individual area of expertise. Much of this was driven by the need to sanitize living areas and to allow for proper hygiene in the cities.

Structural steel allowed for the construction of multistory buildings, including the Empire State Building in New York City in 1930. It was one of the first fast-track jobs that took place in the construction industry. The work moved forward at an incredible speed as they averaged nearly a floor a week setting steel. They even had a railroad track installed through the city to allow for the steel to be delivered directly to the site. It was said that the steel arrived on site still warm to the touch. This allowed for the project to be completed in under one year.

We look at generational changes in the work that is conducted today. We *work* differently today, and *approach* work very differently today. This is a look at the generations starting during the 1880s through today.

Generations on the Job

"Have to Work" Generation

This generation was the first generation on the job as work was organized and the first unions made an appearance in the industry. The sheet metal union started out as the Tin, Sheet Iron and Cornice Workers International Association. The first convention was held on January 25, 1888, in Toledo, Ohio. Delegations representing the original seven local unions were in attendance. They came from Peoria, Illinois; Kansas City, Missouri; Omaha, Nebraska; Memphis, Tennessee; Youngstown, Ohio; Toledo, Ohio; and Dayton, Ohio.

Work, in general, was hard and mostly unsafe. Many, including children, were still working in very dangerous jobs, and no safety concerns were addressed. If you did not work, you did not eat. No social programs were in place to assist workers or families. The working poor remained the working poor.

Vacations were only for the wealthy and did not include any of the workers. Workers mostly labored ten hours a day, six days a week. No overtime or any benefits were offered. If you were to die on the job, it was the wife's duty to remarry someone with a job to provide for the family. Work was hard and little wealth was to be made from manual labor.

"Need to Work" Generation

This generation basically did not have any wealth passed on to it from the previous generation except for an excellent work ethic. The generation showed up for work every day and never expected anything to be given to them. Organized labor fought for and won the forty-hour work week. No safety concerns were yet to be addressed. The work was not plentiful at the time, and many needed to work but could not find any jobs. If you did find a job, you worked as hard as you were able to keep the job.

"Want to work" Generation

This generation served in World War Two and enjoyed great economic opportunities during the post-war times. Work was plentiful and this group only wanted to go to work and take care of their family. They expected nothing in return, only their personal safety from foreign powers. They were a very brave generation and did not take vacations or any time off. They just worked. This was my father-in-law who worked his entire life to better provide for his family.

Forty hours were given each and every week. And no time off.

Work came first and they served their families by providing for them.

"The Baby Boomer" Generation

This is my generation. This generation saw the previous generation work hard and get rewarded for their efforts. They also worked hard and came to work every day. But they *did* expect to take a much-needed vacation and worked hard to achieve that time off.

This generation saw the fruits of their labor pay off, as well.

"X" Generation

This generation entered the job site with a little different expectation. They felt that family was important enough for them to be at every one of their children's sporting events and would miss work to be with family. They also expected longer vacation time and planned accordingly to take whatever time that was necessary to be with the family.

This was the first generation that missed work with good reason to be with family. Work was still important but was not the only thing in their lives.

"Y" Generation

This generation is called the Y generation because it is derived from the question: Why? Why do you think that family is more important than the job? The extended family had prospered, and the recipients of the good times benefited from the previous generation that did not miss work. This group is happy to miss work with the expectation that they deserve to have a nice truck, nice house, and long vacations. They were, at times, given the money to make all of this happen without necessarily working for it.

Working was okay but not their biggest priority. This generation could not be threatened with the loss of work to get them to be on time and stay for forty hours.

"Z" Generation

Generation Z, the most current generation, is now hitting the jobsite. Again, a group that has high expectations and will go deep into debt to get what they want. This is in direct opposition to the previous generations who would only buy what they could afford at that particular time.

Motivation versus Manipulation

The first generations of workers on the job, the "have to work," "need to work," and "want to work," could be easily manipulated into working by threatening to fire them; however, the next generations did not respond to this threat in any way. They would lose nothing as family would step in to make sure they prospered.

Motivation became the only way to get this group to *want* to work, which became a problem for those in supervision. This was a major shift in the way the worker responded to the efforts to get the workers to see the importance of working to the goals that were set in place to achieve the necessary results. It became necessary to get the worker to *want* to work and not be *forced* to work. This was a major change in the process of working.

The "Z" generation is very used to seeing the clock run down to zero as the game ends. They see goals in a way that previous generations did not see them. They are used to goals being laid out in front of them. They are great at being made to think that the goal of making money is a way of motivating them to *want* to work. They will not be *threatened* with forcing them to go to work.

Chapter Three

Products and Procedures Observed

I have had a wonderful career. I was placed in a variety of locations where I watched a tremendous number of products manufactured. I lived in southwest Ohio, in an area known as the Miami Valley that was full of innovative people. The airplane came out of Dayton; refrigeration came out of Dayton; the electric self-starter; the step ladder; the first bar codes; the pop top can and many more items of practical use. Dayton was a place of innovation, and I watched many of these products, and more, being produced. I saw potato chips made, steel poured, glass melted, magazines printed, cars and trucks assembled, car parts made, and tested, semi-trucks and army tanks assembled, and space capsule heat shields treated. I worked on square ductwork being fabricated, triangle aluminum duct constructed, fire trucks outfitted with new sheet metal bodies, and just about anything out of metal coming to life.

I had the honor of moving out of the office position to return to the shop to fabricate an item that no one else thought they could make. I stopped answering the phone on my field supervising duties long enough to make the piece from a heavy gauge metal. It ultimately became the throat of a steel furnace. The piece started out as an ellipse and then transitioned to an oval and then transitioned to a rectangle. No computer program was available to assist in the making of this piece. It was hand-crafted, made from scratch out of flat sheets of metal.

Many items in the shop were fabricated and some were truly works of art. On one occasion we were to construct burglar bars to be mounted into ducts in a prison. Each run of ductwork that penetrates a wall in a secure area requires a set of burglar bars. These bars are in place to prevent someone from crawling through the duct to escape. A welded steel frame surrounds the bars with steel plumbing pipe to form a barrier in the duct. Each pipe contains a solid steel bar that is fitted loosely in the bar. If you try to cut the pipe, the bar spins around and it is impossible to cut.

We would fabricate these as small as six-inch by six-inch bars to fit into the duct. That would be impossible to crawl through. The bars were to be heat-treated to a Brinell 400 hardness. The heat treat company had never heard of something so hard. It would prove to be so brittle that if you dropped a metal bar, it would shatter.

Another unique experience I had involved the making of latex gloves. We were involved in the job of making protective guards for moving equipment. Try to imagine castings of a hand, made from aluminum, from the wrist up to the end of the finger. Now imagine ten of these castings fixed to a conveyor belt with row after row of hands.

They all appeared to be waving at you as they moved along the belt until they were lowered into the liquid latex and clung to the mold. The mold moved through the oven and dried in place. When the mold exited the oven, it was cured. The hands then traveled to the end of the loop where air was pumped into the mold, which blew the glove off each mold. Once the glove was blown off the mold, the process started over from the beginning. Watching this process was certainly entertaining.

The next project was to install a shot blast booth at a company that poured very large castings of metal used to build power plants for ocean-going vessels. The casting was cast in a very rough form with all the warts and bumps and surface blemishes. Next, the casting was moved into a shot blast booth. The booth used a blast operation like a pressure washer used to clean everyday items. This differed as this process used dry shot blast manufactured using apricot shells or similar blasting materials. The booth was constructed of sheet metal, so the work fell upon the sheet metal worker to assemble it with parts supplied by the manufacturer. Often the manufacturer would send out a manufacturer's representative to oversee the installation. We started the project without any difficulties and progressed well through the installation. When the rep arrived at the job, he mentioned that he had just arrived back in the United States after working on a project with a similar booth in East Germany.

Our project was moving quite fast, and he did not need to answer many questions. He replied that this project was quite different from his trip to the Russian-controlled Germany. He stated that the German government was building a plant to produce trucks. They had worked on the building and production line for nearly *ten* years and were yet to produce their first truck. He left the job as the government was in the process of giving up on ever completing the building or production. He had moved back and forth between East and West Germany and spent some time talking to the crew about the distinct differences between the two halves of the same town. Even the colors of the houses, gardens, and street-side flowers were starkly different.

He was at our job for a short time and left feeling that we had the project in good hands and expected to complete the job early.

One snag we encountered after he left was much to my surprise. I was discussing our job site with a representative of the Operators Union as he was trying to get us to put on a tow motor operator for use on the job. An operator would be needed for a very short time and would have limited responsibilities; I felt we did not need an operator on site. He strongly disagreed.

We were well within the terms of our contract and did not plan to add the extra manpower. As I continued to discuss this matter with the Operators union business rep, I noticed our apprentice standing patiently off to the side, waiting to talk with me. I finally excused myself from the conversation and asked the apprentice what he needed.

He informed me that there was a *fire*. I quickly asked if they had tried to put it out. The response came back that they had emptied several fire extinguishers and then tried extinguishing the fire with a 4" fire hose and were not having any success in putting the fire out. He was waiting to ask me if it was okay to call the fire department.

Needless to say, I ran quickly out the door and saw flames soaring into the air. A call to the fire department resulted in the engines quickly responding to the site. We had been cutting apart the old shot blast booth in the scrap yard that was situated over a gravel area that was filled with old oil, which had soaked into the gravel. The fire had a pretty good start as it was burning off scrap oil and whatever else was in the scrap area. There were several sealed liquid drums, and they were starting to swell into big round metal balls, ready to explode. I met the firemen as they moved in to fight the fire. They were not concerned about the potential that the drums might explode. *I* sure was concerned. Fortunately, they successfully fought the fire and moved on to another day. We completed our work, as well.

As the contractor that I worked for was small, I normally served as a field superintendent, supervising the working crews, but at times would take on the role of project manager. As a project manager, I often supervised more than one trade. Our job included mechanical responsibilities as well as sheet metal tasks; my contractor provided both trades, so I managed both plumbers/pipe fitters and sheet metal workers. On this job, we had to work with a general contractor who was less than cooperative. He was known to do things his way and no other way.

A scaffold tower was in the middle of a large room and had been unused for an extended period. Our fitters had asked on several occasions when the scaffold could be removed. It was located directly in the way of the process piping. The general contractor was not assisting or working with us, much to our aggravation. As our project manager, it was time for me to become involved. I asked when he could move the scaffold and was blown off, like usual.

It was time for action. I went to the project engineer who ran the weekly progress meetings and really didn't want to sound like a crybaby. I told the engineer rather discreetly what we were planning. Our fitter foreman dreamed up the perfect solution. We had six process piping pipelines to be run where the scaffold was sitting.

During the weekly meeting, our fitters promptly dry fitted all six lines together. And ran them *through the scaffold.* Dry fitting means that there was no permanent joining of the pipes, but they appeared to be installed with nothing to do but cut the pipes out or cut the scaffold apart to remove the pipe. Scaffold comes apart two ways, and the pipes were run so the only way to remove the scaffold was to cut the scaffold apart.

The plan worked to perfection as the general foreman, who was not cooperating, was attending the meeting with the plant engineer and myself while the fitters worked to complete the plan. When the meeting broke up, I walked back to the job while delaying the engineer long enough for the foreman to get back to the job and see his scaffold held hostage. The general contractor was irate as he quickly saw what had happened during the time of the meeting. Screaming accompanied his tirade.

He quickly moved into the hallway to complain very loudly to the engineer, who, being in on the move, calmly listened to the complaints. While the general contractor was loudly complaining, the fitters were moving quickly to disassemble the temporary piping. When the engineer, the general contractor, and I finally got back onto the job site, all was removed.

The general contractor entered the room and started to point and wave at the horrendous action taken by the fitters, but all that stood in place was the scaffold. The general foreman looked like a raving idiot in front of everyone. The plan had worked to perfection. His cooperation increased from that day forward.

Working in a nuclear facility can be an unnerving day on the job. I had the opportunity to visit a facility that had closed many years before, but was never decommissioned as it should have been. The government Superfund had made money available to do the cleanup of the plant. We were walking through the project with many contractors represented in the group, moving through the facility to see where the work was to be performed. The fuel rods were all laying around on open wooden skids sitting on the asphalt pavement.

As we moved through the open aisles, we were told that if the spent fuel rods had a tag attached, then it meant that the rod was spent and thereby safe to move about. Many did not have any tags attached *at all.*

As we moved down an aisle, several of the tags were blowing in the wind down the walkway. I asked if we could safely say that those tags should be fixed to a rod. I was told very cautiously that all the rods were accounted for. We all could see that they were not. We were also instructed that we would all have to wear name badges and tags that indicated we had been monitored for exposure to radium. Each day we were to turn in the badge for scanning. We were also instructed that we would need to change clothes upon entering, change clothes upon exiting and shower before leaving the site.

Furthermore, we were to be escorted by plant security during the workday as our workers were never to be left alone. This would require us to wait each day for an escort before we could start the workday. At times, this could be a *significant* delay.

After all these instructions, I chose to return to the office and decided not to bid this project. Sometimes the best choice was not to bid a job. An interesting day on the job, no doubt about this one.

Most of the contractors I worked for early in my career did a great deal of work in foundries. Dayton had many steel, aluminum, and other alloy metal foundries that supplied many products made in their operations. Some castings were tiny while others were as large as a tractor trailer truck. Working in a foundry often included a trip to the sand tower. The sand tower was present in most foundries as it was used to move sand up to an elevation that could be gravity-fed into different areas of the foundry to put sand into the molds. The sand towers often were sixty to eighty feet tall.

As an apprentice, I was chosen to go up into a tower, move through the opening, and hang outside to secure a louver on the upper portion of the tower. The crew had shoved a wooden four by four out of the top of the tower and wrapped a rope around the wood. They then placed a come-a-long or rachet device with a long cable attached to the rope. It was then my turn to be ratcheted down about twenty feet into place to attach the louver. I was secured to the rope using a safety harness. None of this would be allowed today due to safety concerns.

I was hanging about 80 feet above the roof below. It was also a spectacular view of the city. Everything was going well until the crew remembered that they had forgotten to bring the screws with them to secure the louver. I had to wait patiently for a crew member to climb down 80 feet and back up 80 feet to bring me the screws. I was in a harness that wrapped around my legs and cut off their circulation as time went on.

Again, all of this happened well before OSHA came into play. And I was an expendable apprentice. But it was a spectacular view of the downtown. For many years, I often traveled by the foundry, looking up in amazement and remembering the placement of the louver in the tower. Another one of the "***I built that***" moments as I traveled around my hometown.

Most of my early work was for contractors that designed and built round duct for dust collectors. Later I worked on a great deal of square duct jobs that provided heating and cooling for occupied buildings. There was only one time that I worked on triangle duct.

It was on an interesting project that the architect chose aluminum triangle duct over the normal square duct most likely for the visual appearance. The job was at an area school, and I had gone to work at the school as a crew member, working for a friend of mine. I was installing the duct and VAV (variable air volume) boxes in the elementary wing, and we were staging all our unopened boxes of grilles, registers, and diffusers in the unfinished gymnasium.

During the night, a student broke into the building and set fire to the stack of boxes. The fire was hot enough to trigger the fire dampers at the 185-degree fusible link setting at EIGHT feet. That means that the fire was over 185 degrees at six feet off the floor. The fire swelled out the top of the gym and caused stress cracks in all four corners of the building as the building expanded due to the high heat. It also melted our exhaust fan blades made of aluminum in the center of the gym. It was a hot mess.

The building had to be certified as safe to enter by the structural engineers before we could go back into the building. They did so and we went back to work to repair the damage.

Our local school district visits this gym each year to play basketball, and every year I have the opportunity to retell the story of the fire. It is also a time for me to look up and see the triangle duct that was fabricated in our shop under my supervision (as I moved into the shop as shop foreman before the new school building opened.) Moving into and out of supervision was a part of my career as well.

The building was completed, and the opening of the building was planned for a very warm day in the fall. I planned to attend the grand opening as our company representative accompanied by my wife who was serving on our board of education and planning to build a new school in the near future.

So off to the dedication we went. One problem existed that was quite unexpected. The building had no air conditioning on this important day as the grand opening was held on a Sunday, and the building was normally unoccupied. We had fabricated and installed the project but were not in charge of the mechanical systems operation. The control representative did not attend, and no one had instructed the owners how to turn on the system manually. How embarrassing.

Later, when the school opened, the system failed again. The controls did not operate properly *after* the control engineering company was let go as a cost-saving measure by the school district. Again, it was not our fault, but it was never a good thing to have a system fail to provide proper heating and cooling.

In our neighboring sheet metal union's jurisdiction is a Honda assembly plant just outside of our local area. The plant is located in the Columbus territory, which meant I received travel pay to and from the jobsite--about an hour and a half drive each direction.

The project was to install sheet metal siding on a building expansion that was taking place at the assembly plant. It was a shrewd move on the customers' part as the state was providing money for the expansion of the building.

The job required us to place new siding on the newly installed steel and to remove the existing siding and store it for a year in trailers. The process was repeated yearly for four years as new assembly lines were added to the plant. Every year the plant was expanded by eighty feet as it added another assembly line.

We were working along, making good time on the project. I noticed that our estimator was coming to the job one afternoon, and I saw him enter the parking lot. The estimator is the one responsible for estimating the project using historical data to determine how long a task would take to complete.

We were installing siding on the building. Our view was great from forty feet up. I also noticed that the estimator had parked in the far lot and then was using binoculars to watch us work. That was rather irritating. So, I had the entire crew of six workers on the count of three all turn and wave at the estimator. This somewhat irritated him.

He came on the job and was quick to comment that we should not treat him like that. He further informed me that he was no longer going to honor our contract and cut my travel pay to zero.

My response was very simple. I started to put my tools in my work bucket with the intention of going to my truck for the long ride home. I told him I had keys to my truck and knew how to use them. I continued putting my tools in the bucket, but he was taking my tools back out. It was a race; I was winning.

I pointed out to him that my crew was putting up more siding than he had estimated, so we were making money. Why he needed to mess with me was not clear to me. He quickly relented and promised to restore my travel pay.

On an earlier visit this estimator was difficult to understand. Even though we got along quite well, he would bring up things just to be annoying. At one point he informed me that the screws we were using cost a quarter apiece. We knew that the screws were expensive, but certainly a necessary part of the job. As he crawled over to the roof edge, he was observing very closely what I was doing. So, in response to this, I began sorting through the screws in my apron and began throwing a few of the screws to the ground. After he looked at me as if to question this, I informed him that I was throwing out all the screws with the head on the wrong end. He was less than impressed.

As we were installing our siding on the upper part of the building using a low roof to access our location, we came across a cable that ran out of the building and across the low roof and down to a small outbuilding along the side of an access road. We did not know where the cable came from or what its intended use could be. All we knew was that it was in a dangerous location, laying on the roof of the plant.

We had cautiously worked around this electric cable, but then the crew decided to unplug the cable on the inside of the building. Safety first.

We continued to work to install more siding. As we worked, we noticed that the general contractor that we were working with quickly visited our site with no communication to us. Next, we saw the project manager make a quick visit, and then the plant engineer came up on the roof. Something was up.

Finally, all three got together and moved our way. No more cars were being assembled as the plant was dark. Apparently, the cable that we had unplugged was quite important, as it powered the entire addition to the building.

The plant engineers and general contractors were quickly moving around the roof. It was then that they finally figured out that *we* had cut the power to the building. A fast cut through the wall and the reinstallation of the cable brought the plant back to life. Who could have guessed how important that electric cable was to the plant and its operations?

Working at high elevations is a part of construction. Working high above the making of steel is a normal part of a day's work at our local steel mill. It is one of the largest steel mills still in operation in the U.S. Recently the mill announced an upgrade that is valued at $1.8 billion. Watching the production of steel being shaped into automobile steel sheets and other steel products is fascinating.

During a plant shutdown, we were working about eighty feet above the mill floor. A shutdown means that you only have a limited amount of time to complete your job and then the mill goes back into operation. We were to have 48 hours before the production line came back to life. While we were moving along well, the decision was made to cut our available work time by four hours and give us only forty-four hours to complete the job.

And safety was a concern for this job. We would always strive to work safely and often have a safety officer assigned to our crew. The safety officer, at times, did not work well with us. On this job, we needed to safely secure a ladder into position high in the steel in the steel mill, and the mill had assigned us a safety officer.

I prepared to go to the top of the ladder to secure the ladder in place. As I attempted to climb the ladder to secure it, I was stopped by our safety man. He instructed me that the ladder needed to be *secured* before I climbed it to *secure* it. What??

I finally distracted the safety officer long enough to *sneak* up to the top of the ladder and secure it. Problem solved. Until I needed to unsecure it. Fortunately, he was elsewhere when it came time to remove the ladder. Oh, the games we play.

This safety officer was also monitoring our use of a gas welding machine. We were working at the eighty-foot level on a platform, with thirty feet to the roof, in the air, and not at all in an enclosed area. We were doing just fine on this very aggressive and dangerous job with a limited amount of time until the safety man stuck his monitor into the exhaust of the welding machine. Of course it failed at that point. When monitored correctly at a reasonable distance from the machine, it passed with no problems. All this occurs when you are on a tight schedule with not much room for error.

From our high vantage point on the platform, we were a witness to a line failure that sent the coil sheet shooting into the air when it failed. That was interesting and very scary. That is also why there is no one allowed in the production area when the steel line is moving. The steel mill can be, and often is, a very dangerous place to work.

The steel mill also reminds me of a time I worked in an auto plant that used robots for the creation of the hood of a truck. The operation was supposed to be in place where the hood is put together with spot welds on both sides of the hood. We watched as the spot welders did not let go of the hood on one side. The robots threw the hood into the air, and it landed twenty feet away. Again, interesting, but very dangerous. All the equipment was well guarded for that exact reason. After final adjustments of the spot welders were dialed in, the plant ran to perfection.

I also watched one year as the final assembly line of S-10 trucks was making for the completion of the paint job on each truck. A human painted the truck with a paint sprayer and the robot would then duplicate the efforts of the human on the subsequent line of trucks to complete the work for the production year.

This initial truck was completely painted right up until the finished product. But the operator had missed one little spot at the very end, so the operator returned to the truck to make one final little spray to complete the paint job.

It was quite interesting to walk past the final paint line and see that little effort at the end repeated over and over by the robot as each truck went through that process for the entire next year.

As I moved forward in my career, the opportunity presented itself to me to watch the full-size pickup truck being built in the Fort Wayne Assembly plant. This offered similar oddities as the final work was completed.

Our job was to build an enclosure at the end of the EL-PO line, the Electro-Plate Phosphate line. This was just prior to the final painting of the truck. Absolutely no silicone could be present in the process as paint does not stick to silicone. We were working on the model of the truck that someone at the plant had really messed up by not filing the paperwork for the emissions control. The engineers in charge of the project did not get the final emissions paperwork in place to formally *sell* the truck. It had something to do with the exhaust emissions during the middle 1980s. So, the truck was being produced but could not be sold. They had produced enough trucks that every plant in the world would receive trucks for plant security but could not be used on the open road. The rest went straight to the shredder right from the assembly plant. Each truck was complete with spare tires, radios, and all accessories. The company was finally able to produce a truck, which was not a full year of production, but a half year for this year only.

We also witnessed the truck frames being fully dipped in the cleaning process before primer coat and final paint. The primer coat of paint was applied during the electro plating process. One evening during a normal run of trucks, we watched the plant engineers sprinting through the plant. We could not figure out what was happening. We finally discovered that the chain-driven line had failed, and a truck was stopped in the tank. The longer the truck was in the tank, the thicker the primer would be applied.

After a full primer coat, the production line finally restarted, and the truck frame moved out of the tank. The truck frame had a primer coat that was about an inch thick, and every assembly hole was closed off to use, so it was very quickly scrapped. Sometimes things do not go as planned.

Sometimes what we are asked to do when approaching a project to be completed does not make sense. Many times, not much is shared as you move forward to completing a day's work. Traveling to Springfield, Ohio, for a small job had the crew working to temporarily remove part of the roof in the factory so that another contractor could work in the plant replacing machinery. We were to remove a large section of the metal roof for the millwrights to crane lift a large stamping press into place.

Later that day we were to replace the roof section. We were all curious how this was to be completed. The large press went into a deep pit and would not fit under the building steel. The hole in the floor was just big enough for the machine to fit. Certainly, a challenging project. The mounting studs for the press attachment to the floor were in place. It certainly appeared that the room necessary for the workers to be under and around the machine did not exist. We were all wondering how this was to be accomplished.

The next move for the millwrights was to place ice in the hole and set the press into place on top of the ice. As the ice melted, the machine was lowered slowly into place. The workers guided it down ever so slowly. A pump was employed to remove the water as the ice melted. It was a slow process, but it worked quite well. As the press was being slowly put into place it was our task to replace the roof that we had removed so the crane lift could be completed. Watching other skilled trades work to accomplish a difficult task was a perk to the job that was quite satisfying.

Coordinating with other trades is one of the items that requires clear communication between the trades. There are only two trades that require drawings to be submitted. One of those is the sprinkler fitters as protection from fire requires the designers to comply with state regulations for sprinkler head locations. Life safety is obviously a major concern while constructing a building. Sprinkler companies must comply with many regulations for the building to be safely occupied.

The other trade that requires drawings is the sheet metal trade. This industry produces a drawing because the trade is full of options as to the way you can design a job. The sheet metal contractor is the only trade who fabricates what is installed. All other crafts can order their material from a wholesaler. With the drawing in hand, sheet metal contractors have the details on what and how to fabricate based on industry standards.

It is customary for all skilled trades to work from the sheet metal drawings. The sheet metal contractor customarily plays a part in running the coordination meeting with the other trades as everyone is working with the sheet metal drawings.

At times it was necessary to work a distance from home. While running a project in Cincinnati, it was also time for a major cancer research facility to begin in Columbus at The Ohio State University. The contractor where I was working was successful in bidding the job to complete the ductwork for this major project. The building construction had fallen behind schedule by several months even before we arrived on site. The coordination of the project normally takes place prior to the construction start.

The first issue was the concrete pillars to be placed under the building that supports the structure. As I worked on coordination for the project, the delays started to impact the completion of the job. The engineer somehow missed or underestimated the impact of a cavern below the work site. As the pillar contractor was moving along to place the pillars, he was dumping and dumping and dumping concrete into place, using many more trucks and yards of concrete for which he did not estimate.

He eventually ran out of money on the project and simply packed up and left. We were notified at the weekly progress meeting that he had gone, and the owner was unsure which route to take to move toward project completion.

As the delays mounted, the basement was finally moving forward and the concrete walls that enclosed the x-ray equipment were put into place. It was then that the owner requested the walls be moved so new equipment could be put into place as the equipment that had been designed for use had become obsolete before it was even installed.

The look on the face of the concrete contractor was unforgettable. Moving or changing several feet thick concrete walls was not something he had even considered. This was a major change.

This issue was taking place about the time that another issue became apparent. The engineering concept drawing that showed where duct was to be installed was found to be in error as the duct would not fit as drawn. The structural concrete supports in all the hallways were not calculated into the process and were too low to accommodate the utilities. This included our ductwork.

As we moved through the options of lowering the ceiling, either moving the duct into the patient rooms or resizing the duct was being explored. None of these options were acceptable according to the engineer.

Lowering the ceiling was out because the ceiling height had to meet the building code; moving the duct into the patient rooms was out due to the many wall penetrations that had to be made; and resizing the duct was out as the units had already been ordered and the order could not be cancelled.

The engineer's answer was to promptly get up from the meeting and leave. And he did. In the process of constructing a building, it is the job of the engineer to offer solutions to a given problem. Leaving without an answer is not an acceptable option. This prompted much written concern from each contractor.

Eventually the project was completed and is now a world class cancer research and treatment facility called The James, but not without a real struggle to complete. I did not have the opportunity to stay on this project as it was completed. But I did see the project start on its journey to completion by others.

As I moved through my career in the industry as an estimator and project manager, I was not screamed at often, but the times I did get screamed at were certainly memorable. One time was during the ventilation of a restroom at a local General Motors plant. It was a small job, and we were subcontracting to a local general contractor. I had quoted the job and had included the line: "We do not include any insulation or installation thereof" as the general contractor had other insulation that he should include with the quoted number.

I thought that my statement was quite clear in the estimate. We were awarded the job as we were the apparent low bidder. Sometime after, as we were moving forward with the job, the general contractor somehow was made aware of the insulation that was not included.

This prompted a call to me. He was screaming in the phone telling me that I *would* include the insulation in our price. I very unemotionally informed him we would *not* include it in the job. Screaming by the general contractor continued. When I presented this information to my owner, he was equally firm on my not including the insulation in the price. My supervisor followed up with a phone call to reaffirm our intention of not including the insulation or installation in our price. I can still hear the general contractor screaming at me.

Another time of full-blown emotion included the time I was called to the General Motors truck assembly plant. We had been contracted to provide spot cooling on the assembly line. This means we were to run ductwork along the final assembly line to each worker's individual area of work. The work was to be done as the assembly line was in operation. We could safely perform this work as we moved over the line. An angry call was made to our shop, but it was made to our estimator who was on vacation. It was my role, as the field superintendent, to go to the jobsite and fill in for the estimator. As I approached the job, I went directly to our jobsite foreman who quickly told me that our workers did not stop the line. Just like our workers had been instructed.

I then moved on to the engineer and listened as he was quite upset. The line produced a truck every 58 seconds. The line was shut down for eight minutes, which resulted in a loss of about $10,000 per minute for the customer. I didn't think that my writing a check for $80,000 was a very good option, so we moved on to who really had shut the line down.

Our foreman identified the individual who had pushed the stop button. The engineer approached his line worker and asked if he had, indeed, stopped the line. He admitted he had done it on his own because he thought our crew was at risk.

We were not at fault, but nothing else was said by the plant engineer. No "sorry for getting you here". No "sorry for the *plant* shutting the line down". *Nothing said at all.* I left for the shop.

The next time that I was the target of someone yelling was not meant for me as I was filling in for someone yet again. This time the call came to our shop when I was covering for someone who had estimated a job that was to be fabricated by a supplier of plastic duct work. The project required ductwork to be made of a material that held up to caustic fumes and vapor that would destroy metal in a short time.

The duct manufacturer was to supply all the necessary material for the entire system of ventilation over a dip tank. The materials would come from the factory built in pieces, and it was our job to install all the pieces to complete the system. It would require us to cut, fit, and weld all the factory-provided material. The skills were identical to making it from metal, but the system was to be made of plastic.

The angry call came from the local GM plant that did not offer us much work. Another contractor had an "*in*", so we could not seem to move into the plant. This call originated directly from the purchasing department. This was not a normal call. When the call came in, I was filling in for the owner who was out of town. I grabbed the file and headed off to visit the very irate customer.

The purchasing agent immediately informed me that I must have made a mistake on the estimate and wanted me to drop out of the running for the job. It was a nice project and would earn our company a good sum of money. I quickly reviewed the estimate, and it all looked good to me. I informed him that we would stand behind our numbers.

In the background was the plant engineer who I had worked with on several previous occasions for other contractors. He quietly sat in the background and did not enter the exchange. As the conversation moved forward, it got very loud, and I was not the one screaming.

It seems that the purchasing agent had already guaranteed the job to his friend, the other contractor who had the "in", and he felt strongly that we could not do the job as cheaply as we had bid the work. I left the office after, again, clearly standing behind our price.

After my owner returned to town, I quizzed him on the situation and was met with nothing more than a smile. I knew something was amiss but could not figure it out. Nothing else was said until the purchase order arrived by mail.

Then I was let in on what was happening. The job was fixed. The salesman for the plastic duct company was in on the take. The purchasing agent was in on the take. The other contractor was in on the take. *I was not*. My owner informed me that he had gone directly to the company for a price, leaving out the crooked salesman. I was instructed to see that the job was done to the best of my abilities. That is exactly what I did. The job went well. The engineer was pleased.

Meanwhile back at the factory, the crap was hitting the fan. The salesman was called in and fired. The other contractor was in a mess, as well. The purchasing agent was also on the block as he did business this way too often. He was reassigned to another plant run by GM. Far away. He was also buying and selling cars with his discount used under several different names. Also, quite against the rules. No wonder we could not get work at that plant.

Through the years, I worked with many talented sheet metal workers, sometimes referred to as "tinners". Numerous workers were very good at their jobs. Some overcame challenges that many would find difficult, if not impossible, to move on to a successful career. I was afforded the opportunity to learn from and learn about a very diverse workforce. The late sixties was a time of great need for building trade workers. Many of the sheet metal men came from parts of Kentucky and Tennessee, and most came into the industry through family. All had served the required apprenticeship that was the normal route to entering the building trades.

As an apprentice, I mostly worked in the shop. This time I was working second shift. We did not do much second shift work. As the week progressed, one of my jobs was to assist the foreman with filling out timecards, which helped with cost accounting, as the men moved from job to job.

Because we were working the second shift, it was necessary for us to predict on Friday what we would be working to complete, even before we did the work. I was helping the part-time workers fill out the timecards, and one worker had a panicked look on his face as I asked for his complete timecard.

After some time and effort, I asked directly for the card. I was told, with much embarrassment, that he could not do that. He would take the card home every day and along with it the information to have his wife fill out the card. He had done that several weeks in a row without my knowledge.

The worker then shared that he could not read or write anything at all. This came as a complete surprise to me. He had hidden his secret very well until I pressed him for the timecard. I just could not understand how anyone could function on the job as well as he was able without the skills of reading and writing. But he did. I felt terrible that I had pressed him into that revelation.

Over the years, I have met and worked with many highly qualified workers in a variety of trades. Nearly all tradesmen are good at what they are paid to do. One individual was a crane operator who was very skilled at operating a crane; he controlled and moved the load with great skill at just the right speed to make working with him a real pleasure. As you work with an operator, your fingers and hands are in a very risky place as the heavy load moves into its final resting place. I fully trusted him.

This same operator was called to a job with us that required him to lift a dust collector out of a courtyard one floor below ground level. We were to remove all the connections to the unit so the lift could take place. Often it is a safety requirement that the building be evacuated during the lift, which took place on this job. All the workers from inside the building were moved outside, mostly sitting on parking blocks in the parking lot. I positioned myself next to the crane and the operator. As the line on the crane tightened, it was time for the lift to start. Up went the dust collector complete with dust on the top of the unit. The unit quickly cleared the power lines and was moving toward the flatbed truck that was waiting in the parking lot. All looked good.

This was when the load on the crane slowed to a stop at the top of the lift. The operator looked at all of those on the curb and then looked at me. He asked if I wanted to have some fun. I was game.

The operator expertly moved the load to the ground and started to tip the unit over to lay it down. He then started his move. He shook the line, which resulted in a cloud of dust falling from the top of the crane load. He then started screaming, jumped from the crane and started to run across the lot. As soon as this happened, all the workers jumped up and started running for their lives. Almost instantly, the operator stopped, looked at me, and started laughing as all the parking lot runners turned to see him howling. Just another day on the job.

Operators that are in total control of their rig are a must. There are many good ones. This one just really enjoyed his job.

Another operator that was working with us had to sit on the street outside our location and could not see what he was lifting. We were on the roof in a courtyard of sorts. We were removing a fan that was in the courtyard at roof level.

The fan was to be cut loose from its mounting frame before the lift and our crew thought the fan had been satisfactorily detached. It had not. As the crane tightened the load, the cable started to unwind. The crane started to bend down as the load continued to tighten.

All at once, the last weld broke loose, and it sounded as if a shotgun had fired. Immediately the crane's load jumped up and started to swing and hit both sides of the building. This allowed the operator to finally gain control of the load. The fan was then safely removed from the roof to the ground.

It was then that the operator left his rig, looking for the job foreman to express his displeasure.

Everyone involved was shocked to be a part of a job gone wrong. No one was hurt, but the operator was rightfully upset.

Operators come in different forms. They operate a variety of types of equipment used daily on the job. This includes trucks, forklifts, mini cranes and other forms of transportation and equipment. Also included are air cranes; these are helicopters that may be big or small. They come as large as the twin prop helicopters that can lift large loads.

Every lift is based on weight, so the lighter the load, the easier it is to complete the job. It is expensive to use a helicopter. They involve a great deal of planning and usually include many lifts to make its use affordable. There is little time between lifts, and the pilot/operator wants to move quickly from one lift to the next.

I planned a rather involved lift of over 150 lifts in one day. The building had to be empty to fly over, so the lift was planned for a Sunday, when the building was not in use. The lifts were numbered and were designed to be completed in a rather logical order moving across the roof. The first lifts were for our largest equipment, which were two large identical HVAC (Heating, Ventilating and Air Conditioning) units. Each weighed close to the maximum lifting capability of the helicopter.

Lift one was planned for one of these big units. It went well, except that the downwash from the helicopter, which was about ninety miles an hour, lifted our plumbing foreman off the ground and pinned him to a fence about two feet off the ground. This big guy was stuck on the fence until the copter left the area, heading for the roof of the plant. Up came the load to be placed on our roof curb that had already been installed on the roof earlier in the week.

When the downwash hits your body, you instantly must brace yourself with one knee forward while waiting for the sudden blast. The entire crew also must put on a chin strap in advance to keep their hardhats in place. The downwash takes your breath away.

We were warned to never put yourself under the load as the pilot has a quick-release button that releases the load in case of a problem. The last thing the pilot wants to do is go down with the load. The safety briefing that starts that day makes this *very* clear.

Bringing the load down slowly takes a great deal of skill, but the pilot makes every attempt to do so. This usually causes quite a stir when the pilot cuts the load loose a couple of inches early, which makes for a hard landing. You really need to be ready for the landing.

The next lift was supposed to be the identical unit for the other end of the building. When the lift was attempted, the signal man on the roof with me looked at me and shook his head *no*. The lift could not be made because the previous lift had been aided by the placement of the now-missing first unit to act as a buffer for the down-blast.

The pilot indicated that we needed to pump out some fuel to assist the aircraft in making the lift. The helicopter crew carried empty 55-gallon drums for just such an occasion. Before we could get to the ground to assist him, the pilot decided to move through some of the other lifts to burn off some fuel.

Re-numbering lifts in the schedule meant that we needed to move people into place to rig up the lift as well as to move people into place to land the pieces to be lifted into place. This also involved securing enough rigging to have one on the piece, one in the air, and one to place on the next piece. Many of these moves were like a well-orchestrated play. Everyone needed to know what was happening, because no down time could be tolerated with the pilot moving quickly from piece to piece and lift to lift. We were to average about four minutes per lift.

There were a great number of lifts required. We needed to plan for more than one day of helicopter lifts, so we used several different helicopter companies, which were located all over the country. We needed to plan on using the helicopter company that was in the vicinity of the job, as we also had to pay for any ferry time or movement of the equipment in place. One company was based in Oregon, so we wanted to schedule them when they were working close by. As it turned out, they were in the area due to the Mississippi River flooding. The company needed to move electrical equipment to higher ground.

The other helicopter companies were in Michigan, so those were easier to schedule around. Some of the helicopter equipment was new and others were much older and were heavily used in other endeavors before working for the construction industry. Several had been used in Vietnam in military uses.

Working with cranes can be a fun but potentially dangerous part of the job. A skilled operator makes for a safe, enjoyable day on the job. One job called for a very large crane as the task at hand was to set a piece of equipment a sizable distance back from the edge of the roof, requiring a 200-foot boom with an extension of fifty more feet. It also required us to lift the equipment over high voltage power lines, adding to the challenge.

It was a clear day and good weather for the lift. As I watched the lift from the ground, I was talking with the estimator and representative of the crane company he was sharing a lift story from almost the same location but with one difference: The weather was not clear but was made on a misty and unpleasant day. Despite the weather, the lift was successful.

As their lift was being made, an OSHA (Occupational Safety and Health Administration) inspector stopped by the site. My lift representative was again on the job to observe. The lift was made by the same Manitowoc crane company. As the load was being lifted past the power lines, the electricity arced from the lines about fifty feet over to the crane. It was quite a sight. This can and does occur under the proper conditions but fortunately does no damage to either the people or equipment.

But for the casual observer, like the OSHA representative, it is quite a shocking sight to see. He jumped out of his skin. Our crane rep, who had experienced this many times, was not shocked to see the event. He simply commented to the OSHA man, "You obviously have never seen us jump-start a Manitowoc before, have you?" A great sense of humor helps most every day.

My career has taken me to many places of creation. Witnessing the production of many manufactured goods has been very interesting. Watching potato chips made from scratch was a real pleasure. Our company was to make welding repairs to the chip fryer between batches. This required watching the operation in process before we made our way to the site. I arrived at the plant and met with the maintenance supervisor and discussed the project. We then proceeded through the plant and watched as potatoes were delivered by the semi-truck load and dumped on the floor, awaiting the move into the washer, then the slicer. The potatoes were spun at high speed and slicers on the side of the washer cut the potatoes as they sped by the cutting knives. The potatoes then made their way to the frying process.

The potato slices were cooked in peanut oil and floated the length of the fryer before being sorted by a unique method of removing the bad chips. There is a copper tube with holes strategically placed to blow bad chips down and out of the stream as the chips move over a gap in the belt.

At this point, the maintenance man reached down and grabbed

some chips off the assembly line. He offered some to me. Fresh warm chips straight out of the fryer. What a treat. This was memorable and even more so now that the factory has shut down.

The chips moved on to the area where they received the ingredients to flavor them into BBQ, honey mustard, or any of the other flavors. Those aromas were pleasant, as well.

The final step was to bag the product and ship to the customer. Another opportunity to observe the process of making a product that is enjoyed by all.

Another food product I witnessed was the production process of Corn Nuts, a snack food sold in convenience and grocery stores. The company was moving into the Ohio area to open a plant to process the product, which is a special corn grown in Ohio. The kernel is larger than most kernels of corn. The processing plant bakes the kernel and packages the product for sale.

Other corn products were big in Ohio as Cargill had a large facility to process corn in the Dayton area. The plant was at one time the only plant processing high fructose corn sweetener for all the Coca-Cola products. The sweetener was shipped to Atlanta to be added to Coke products. The corn plant in Dayton was a busy place during corn harvesting season. Trucks would enter the plant and be unloaded by placing the entire truck on a ramp that lifted the front end of the truck so that the corn could be dumped into the corn chute to start the processing.

This system was later replaced by a truck dump that accelerated the process. A truck could be unloaded in minutes, yet the line to unload was still always very long. The corn was separated into many usable parts and completely used in one way or another. No part of the corn was wasted.

Another area in which I gained a great deal of experience was foundry work. I ended up spending much of my younger years in foundries. Foundry work was intensely busy as I entered the workforce in the early seventies. There were as many as ninety foundries in the Dayton area; today less than five remain. Some cast very large castings and others cast specialty items that were very small. I noticed in several of the foundries' display cases there seemed to be a standard display of a teacup and saucer with a spoon. I thought this was odd to be displayed in nearly all the locations, no matter what product the company produced.

After seeing the teacup several times, I finally garnered the courage to ask why the teacup was displayed in all these showcases. I was told this was the final test for a foundryman: to be able to pour the mold for this sample correctly and properly. They were all very intricate and very impressive. The free-standing spoon was the difficult part of the mold process. Now, even I knew the secret.

On a nice Sunday morning, I was called into work to a foundry location in Dayton. Maintenance had observed a hot spot on the side of the melting pot. It was glowing bright red and very impressive. I was not sure why I was called in, but there I was standing on call and ready to do what I could. Maintenance informed me that if the kettle were to break through, the river of steel would go unchecked until it cooled down. This meant that it would slice off steel columns and be unstoppable until it reached whatever and wherever it was going. Again, I'm not sure why I was there.

The molten steel was transferred into another pot and repoured into another furnace. The situation cleared up. I'm still not sure what I could even have done if I was needed.

In yet another steel making operation, I was working under a steel furnace installing a duct to cool the underside of the furnace. I was told that the steel furnace operator would not pour anything out of the furnace while the apprentice and I were under the furnace, as we had no way out once the furnace started to tip out the load. All at once, however, we noticed that the furnace kettle was tipping. Much to our surprise, we were not in danger in our place under the kettle. It would have been great to know that ahead of time.

Molten steel makes me next think of working around molten glass. We were working at a factory that made fiberglass out of molten glass. We were told that this process could not be stopped, and our work needed to be completed while the ovens were in action. We were installing a temporary stack on the furnace so that the furnace repair company could work on the furnace.

Installing the temporary stack required us to install a new roof opening and telescoping heavy gauge stainless steel duct directly under the roof. The plan was to cut the new furnace opening and then drop our duct down to the furnace to provide ventilation to the outside of the building. All over molten glass.

The first part of the plan went well as we cut the new roof opening and secured the new stack to the roof. The next step was to telescope the new heavy gauge stainless steel duct into place. This involved placing four chain falls into place to lower the bottom half of the stack.

The next step was for the furnace repair crew to cut the opening in the refractory brick that covered the end of the molten glass. This was done by the furnace repair team with the aid of space suit-looking fire protective gear. They used a chain saw equipped with a diamond tipped blade. The value of the blade was over $6000, and it worked to perfection. They mounted a hook into the brick to be removed, cut the opening, and lifted the piece out of the way. A ¾" garden hose was attached to the saw to cool it down. The water evaporated before it even hit anything.

The next step was to lower our piece of duct into place. Four of our workers completed the task. The sleeve began to vent right away. The only issue was that the duct heated to cherry red, and this is where the concern worked on me. I was afraid that the welds would fail, and the piece would fall.

Fortunately, no welds broke, nothing bad happened and the piece slowly returned to the color of stainless steel. We had worked the plan to perfection. The new piece exhausted the furnace as planned.

Much of my early work took place in foundries. Working in a foundry involved working in hot, high, and uncomfortable places. One job was to replace dust collector bags. A dust collector is used to remove unwanted dust, wood chips, dirt, or material from the workplace. It ranks in the top five of the worst jobs on the television show, *Mike Rowe's Dirty Jobs*. Think of a dust collector as a huge shop vac. A shop vac has one filter that is usually a cartridge. The dust collector has rows and rows of bags that run from the top of the unit to the bottom. Usually, it's eight to ten feet tall. The collector might have up to one hundred bags, depending on the size of the collector. They are normally about four inches in diameter. To remove an individual bag, the worker either loosens a hose clamp or another similar attachment. They are dirty, dusty, and foul! Next in the process was to mount new bags in place of the old, dirty bags. This was the easy part. The newer style of collector used today is a cartridge style collector, which again closely resembles a shop vac in that the cloth filters have been replaced with cartridges and are much easier to replace. And cleaner.

The worst-case scenario for a collector is a lamp black collector, which traps lamp black dust. Lamp black is what makes car tires or rubber products black. It is very fine dust and will penetrate clothing or anything else designed to keep you clean.

This was a product found at several of the GM divisions where I worked as an apprentice. It was the worst job you could get to work on. The rubber around the car door and trunk and the rubber used to make windshield wipers came from those plants.

The only time I ever damaged anything on purpose was when I was so frustrated by the construction manager in charge not listening to me. It was not actually one construction manager but a combination of construction managers. There were two different managers in charge of construction; one was a manager in charge of the actual building and one in charge of the interior construction of the project. They did not speak to each other because of an ongoing argument between the two of them. Meanwhile, all of us on the job had no supervision. I had made my building contact aware of my dilemma: I had units coming in that were ten feet wide, but the back hallway would be only eight feet wide if the second wall of the planned hallway was erected. Because the two contractors were not speaking, the second wall was planned to be erected prior to us bringing our units.

The morning of the planned installation was quickly approaching, and the two sides were still not coming together. The planned wall was to be sixteen feet tall, and the base channel had been fixed to the floor. I alerted the carpenters that we were moving forward as instructed about my problem but to no avail. They started their work, and I waited until their morning break to move into position.

As the crew of carpenters took their break, I moved our largest platform lift into position, straddling the base track, and proceeded to move forward, hitting the wall studs and causing them to fall. It only took about four or five studs to hit the floor until everyone came running. I calmly parked the eight by ten-foot lift and took the key.

Now that I had the attention of both sets of construction managers, I could *again* explain why they could not move forward. It was suddenly agreed that the wall would not be erected until I was ready. A little well-placed action and communication goes a long way.

Another foundry where we did a considerable amount of work offered the unique opportunity to move into the world of solar energy. This was in the late seventies, and we were working to design and install a solar energy job. It was designed to preheat the scrap pig iron prior to moving it into the steel furnace. The metal needed to be perfectly dry, and preheating was a benefit. The solar panel manufacturer shared that this would be the first industrial solar job that he was aware of in the country that the company had provided the material necessary to complete the project.

I had never designed a solar project but jumped right into the design. The placement of the panels was engineered to be at a ninety-degree angle on the winter equinox as that was the least effective day for sun, being the shortest day of the year. It was to be placed at a twenty-three-degree angle as, again, it was the worst day of the year for the sun's effectiveness.

We progressed through the design and started building a platform on which to install the panels. The panels became so hot to touch when the sun was out that welding gloves were required to handle the panels. This installation was an air-to-air system, so no water or piping to carry the water was involved. The system worked quite well until the plant eventually closed for business.

The opportunity to work in historic Dayton buildings was also a part of the job. One job was in a building with a rounded ceiling in a part of a GM plant that made steering wheels for many of their automobile products. It is in a series of buildings all built in the same style, the original permanent buildings that the Wright Brothers had built to manufacture the first airplanes for the Wright Brothers Airplane factory.

The airplane area was close to the section of the GM plant that built the first ice cube trays for Frigidaire refrigerators. These were made in Dayton before automatic ice cube dispensers were designed into refrigerators today. The trays were constructed from aluminum and were hand-filled with water; after the water became frozen, you lifted the cube maker to break apart the cubes. Another Dayton invention that I witnessed in person.

As I started my career in the construction industry, I was fortunate to start with a company that treated me with respect and expected a great deal from me. Other apprentices hired by the company were not treated with the expectations of success that I was. I never did figure that out but certainly appreciated it.

In less than a month I was trusted with making my first fitting that would go in the end of a duct run to provide spot cooling for a hot part of a GM facility. That fitting remained there for many years to follow. Upon entering the plant to work in other areas, passing the location many times. It brought pride and great appreciation that the fitting continued to serve its intended purpose. I doubt that no one else shared my feelings.

The Gem City clock tower turret is a unique Dayton landmark. It now is visited by all Dayton area residents and visitors through a trip to Dayton History at Carillon Park. It originally was located at the northeast corner of Third and Main streets on the Callahan Building and later moved to the Reynolds and Reynolds company near present-day Interstate 75 as it passes through Dayton. The tower was later moved to Carillon Park and used for visitors to climb a steel structure and look northward along Interstate 75 through Dayton. In the tower's location along Interstate 75 at the Reynolds and Reynolds building, many Daytonians could look at it while driving past. While working in the plant, the Reynolds and Reynolds maintenance man took me up in the tower. Not a place that is accessible to most people.

As we looked up at the tower's clock mechanisms, we noticed more than several beams of sunlight streaming through the tower. When I asked the maintenance man where all the holes had come from, I was informed that the bullet holes were from guns fired by passing motorists cruising by the clock tower on Interstate 75. The maintenance man said on one occasion, he was even present when the bullets came bouncing around in the tower. An awesome experience while just doing a day's work.

South of Dayton there was a paper mill project with a very aggressive schedule, needing to be very quickly completed. The plan was to work around the clock and assemble a large paper machine while a new building was being built. Normally you construct the building and then install the interior machinery. This project called for these to happen simultaneously. The machine was to produce the thermal paper used to print receipts, such as gas station receipts. The paper is produced on a forty-foot-wide roll and then cut down to three-inch-wide rolls. Another fascinating process to observe the manufacturing of an American-made product. This factory was designed to be built either in the USA or China. The employee-owned company chose to build in the United States. Our part of the job was to install six different units that produced high heat. This incorporated very heavily built ductwork surrounded by thick insulation and then covered with metal cladding or covering. The lightest piece of duct weighed almost 900 pounds.

The heavy ductwork had to be bolted together, and each connection was then covered with insulation and metal. It was installed on an elevated mezzanine and next ducted down to the machine. Much of the ductwork was started at the eighty-foot level and ducted down to the machine connection at floor level. This very heavy duct was installed at a high elevation. This was completed by working two overlapping twelve-hour shifts, so no work slowed or stopped at any time. Our toolboxes closed for only one day, July 4th. Everything else was worked non-stop. This is when you naturally became unbelievably tired on the job. It is a challenge to keep yourself on task and not get hurt. Management told us that if we missed a day of work, we would be fired.

In addition to the aggressive schedule, the construction managers did not want the individual building trades talking to each other. Therefore, the much-required coordination did not take place. But the foremen of the different trades still met unofficially as the job moved forward. Trades are required to work together to accomplish a safe and timely completion.

Because this job was in a paper mill, a "no smoking" rule was required on site. Smokers were required to leave the site and walk out to the main road leading into the site to smoke. It was about a ten-minute walk out to the smoking area, and a ten-minute walk back, with ten minutes of break in between. This actual thirty-minute break was *supposed* to take place in the ten-minute break time. It didn't.

Often, construction workers are treated with little respect. We often are not allowed to use the restrooms of the plant where we are working. We frequently need to park in a temporary parking lot in no man's land. The lot is usually located at the outer reaches of the property. The time it takes to walk into and out of the job is on your own time. Sometimes it was a *very* long walk.

Working with many highly skilled and experienced workers is interesting. Working with not-so-skilled and less experienced workers can be fascinating, as well. I had a trustworthy truck driver that worked for us and was a great worker who was normally on time. One day, to explain why he was late for work, he told the following story: He was locking his door when his dog got out and grabbed his lunch and ran down the road. He chased the dog and realized he was close to the neighborhood eatery. They offered breakfast, so he stopped, which made him late to work. A great story. I told him that a great story was needed for someone coming to work late.

I was late for work about a month later, which was out of character for me as I had not been late for several years. I thought a great story needed to go with this, as well. I told the driver that the house across the street was on fire and that the fire trucks had parked across my driveway, blocking me in for some time. That was my excuse for being late. Next came great concern from several of the girls in the office as each was asking if my neighbors were okay. They were all offering to assist in any way they could. Great story gone bad.

It was also a recommended policy (jokingly) to leave early if you came in late to make up for leaving early.

Later in my career, I moved back into running work as a general foreman. I was working for an out-of-town contractor on a picture tube manufacturing project that was an interesting challenge. At times, it appeared to be a United Nations job as the team of players was quite diverse: the customer was Japanese, the architect was Saudi Arabian, our engineer was Mexican, and the project managers were both Japanese from two competing companies and not overly friendly to each other. The workers from all trades were Americans.

These Japanese firms did work worldwide. At a job meeting, I jokingly referred to one of these companies as a "Mickey Mouse" outfit. The Japanese construction manager asked what my "Mickey Mouse" comment meant. I tried to get him to understand. Later I saw an advertisement for the company in a trade magazine that indicated that they had offices worldwide and realized that they were no Mickey Mouse kind of company. Perceptions are sometimes faulty and embarrassing.

The company where I was employed at the time was based in Kentucky. On a regular job meeting visit, my vice president from Kentucky was in attendance, and I was curious to see what he had noticed to be a challenge for the customer. He had observed that the customer always had one more person in the meeting than we had. We were always outnumbered.

On this occasion, my VP asked me to bring three foremen with me, so we outnumbered the owner. The owner then delayed the meeting as he tried to gather more people. He even brought his janitor into the meeting. But to no avail. We still outnumbered him.

All the while my foremen kept asking why they were in attendance. They were instructed only to listen but not participate in any way.

The owner still could not bring enough people to outnumber us, so he continued to delay the meeting trying to gather more people. Finally, after nearly an hour, my VP told me to send my foremen back to work. I did so and the meeting finally started. Quite comical as this all unfolded. I still don't know what the number of people had to do with that meeting.

Further frustrations continued as the job progressed. It was a continuing challenge to get the necessary information that was needed to proceed. I needed to think through this process that needed to pass through so many people to simply get a required decision. It was beginning to be an unsurmountable situation. I tried to short circuit the problem by suggesting a solution to the problem on a separate sheet of paper with the written RFI (request for information).

This would allow for a faster response to the request as I did not have to wait for an ineffective solution and subsequent re-submission of the request. Our engineer was happy to allow for the solution to be attached to the request.

Our work often takes us places that are not normally visited. I was a first-year apprentice and was working in a plant that made five-gallon metal buckets. I climbed into the overhead steel above an oven and completed my work. Three years later, I was back in the same plant to work again and climbed into the same steel above the same oven. As I climbed into the steel, I found an adjustable wrench laying in the web of the steel. And it had my name on it. I found my long-lost wrench!

A visit to a jobsite can vary as the project moves toward completion. When the army tank plant in Lima was originally being constructed, you entered the job through a guard shack with a guard supplied by Chrysler Corporation, the contractor building the tank. You then entered another security checkpoint where you presented your driver's license and got matched up with an escort into the plant.

I was visiting the site to bid the job of exhausting the smoke and fumes coming off a cutting table, which cut the armor plate that was located on the front of the tank, using a high-pressure water jet. I made my way, with my escort, to observe the process. The armor plate was four inches thick, and the cutting table sliced through the metal to perfection.

At the time, none of the assemblies of the tank were in place, and all that was present was a bare concrete floor. I moved forward and made the necessary estimate for the work.

Later, I returned to the plant when it was in full production. This time, however, no escort was required. As a matter of fact, I could simply walk right into the facility. Which I did. I was able to observe the sixty-ton tank in full production. While working on the roof, we could see the full test track in operation. The test track is a full track with banked corners and looks like a NASCAR track.

The sixty-ton tank was being driven on a banked test track at sixty miles per hour. The tank could stop on a dime. It also would cut through the middle of the test track on its way to the firing range. I watched as the tank became fully involved in what looked like an engine blowing up with smoke everywhere. The plant fire guard, who was working with us as we were welding on the roof, informed us that the tank was indeed throwing the smoke screen down that was a part of its normal operation. You could not see anything for hundreds of feet as the tank was driven through the smoke using infrared by the driver. Really amazing.

We learned that drivers were civil service workers. You could get a job driving a tank around the track. We also asked what would happen if someone tried to take a tank home with him. It was obviously not something you could stop as the tank was one of our fiercest fighting weapons that the U.S. owned. I was told that the plant had a contingency plan for just such a situation.

Another perk of working on a project that most would not have the opportunity to witness.

Chapter Four

The Players Today

Construction Managers vs. General Contractors

The original system of contracting jobs was for the general contractor to work directly for the customer while performing the necessary and required work to complete the project. It was direct contact and there were no layers of management to work through. Progressing through the years, we have added managers, particularly construction managers. The difference between construction managers and general contractors is that the latter formally would take on all the responsibilities of completing the work. They were paid for each task that they, themselves, completed or any of their subcontractors completed; therefore, they had a vested interest in completing the job.

Construction managers, on the other hand, entered the field of construction in the 80s and 90s. They saw a great deal of time spent in court trying to shake out who was right and who was wrong before payment was made to the contractor.

Many construction managers at one time were general contractors but changed to construction managers because they intimately knew the business and would rather settle for the customary 8% fee without the penalties of missing a target date and be penalized for the lack of production due to inclement weather or anything else that might stall work completion.

The new field had no risk and all reward. The customer was convinced that the manager was a necessary addition, who would make sure that the job was done correctly, on time and under budget. This is exactly what the general contractor has successfully done for years. But now the construction manager has taken over and promised the same results with no penalty for not performing just as the general contractor had done for many years.

The main problem with the construction manager is that the manager usually does not have trade experience and becomes one more layer of management on the job. One more entity to work through to get to the customer and potential problems.

Chapter Five

Safety and the Change in Safety

Safety and the emphasis on safety have changed over the years. I predate OSHA. I started in 1972, and OSHA was legislated in 1975 and was enforced in 1976. Before that, we thought that "*OSHA*" was a small town in Wisconsin. Not much regard was given to safety before the law forced the construction industry to focus on how the job needed to be safer. It especially became important when companies became self-insured, and it cost the company much more money to ignore the law. The emphasis on safety was eventually so strong that contractors could not even bid work if their experience rating was too low. Ratings are calculated so that a new contractor needs two years of rating to qualify to bid on work. It was a bottom-line item.

A General Motors facility needed the installation of ventilation for a material produced in a building that was set aside in the parking lot. As apprentices, we were taken to the building and shown the work to be done on the mezzanine level of a one-story building. We were further instructed about the vapor-producing material that was generated in the room. The room was equipped with a fire suppression system that incorporated an extinguisher that basically extinguished the fire by removing all the oxygen in the room within eight minutes.

We were instructed to test our ability to get down out of the building steel, run down the steps, and exit the building within eight minutes. No pressure there.

There was to be a warning horn, flashing lights and blaring siren before the extinguishing chemical discharge--all in the name of GM making their product in the room. We quickly completed the small job and then wondered if the system would indeed work. We did not wish to test it out.

On another project in the same building, during Christmas shutdown, we were to complete a tear-out of the overspray of liquid rubber applied to the outside of the weather-stripping that surrounded the car doors on most GM products. The overspray was a huge hunk of rubber hanging low in the industrial oven used to cure the rubber. It was inside a chamber that included a long section of conveyor that allowed the rubber to dry before leaving the oven. The sprinkler system (water) was turned off as well as the Ansul system (a system that releases a fire suppressing agent). The one that was not turned off was the *deluge system* that dumped 8,000 gallons of water into the oven in a four-minute time limit.

The work was progressing satisfactorily until the end of our 12-hour shift. About 15 minutes before shift change, one of our people accidentally and carelessly waved a torch near the detection head. I was in the oven and near the oven access that was two feet from the floor. I heard the detector "click," which meant it had been tripped. Next came 8,000 gallons of water dumping into the oven. It immediately filled up to the bottom of the door at the two-foot level.

I could not jump through the door quickly enough. I was soaked and did not have extra clothes on the job. It was also Christmas shut down. The temperature outside was a whopping high of 5 degrees that day. All of us were soaked. With water running out of our boots, we were in for a long drying off period. The next shift started while we sat and sat and sat as we waited to dry out so we could go outside to our cars and trucks.

Things were not always better in the warm months either. Working on the roof in the summer months was never good. One July afternoon, we were rebuilding an air handling unit on the roof, and it was blazing hot. We had been working several hours when we all stopped for a short break and sat down in the shade of the unit.

All at once, on a super clear day, out of nowhere, we were hit by a blast of water. It happened twice over the course of the next hour. We had no clue where the water out of the clear blue sky was coming from.

After the last blast of water, we noticed one of the plant guards disappearing down the ladder over the edge of the roof. Mystery solved. Now for payback. We noticed a rain cap laying on the roof. Rain caps or stack caps look like a giant hat that would be worn in the rice fields to protect you from the sun and weather. It would hold many gallons of water. And did. We filled it, took it to the edge of the roof and waited. The next time the guard came to the ladder to climb up, we dumped the water on him. It knocked him off his feet. Repaying that favor was very satisfying.

At times, safety can be confusing and aggravating. On a job in a paper mill, where there was activity in many areas throughout the job, it was a rule that you always install yellow caution tape to indicate an approaching hazard. As I moved about the job, I noticed yellow tape ahead and started to maneuver around the tape and then noticed more tape wrapped around a column and even more tape about ten feet away wrapped around another column.

As I tried to make my way around the hazard, I could not tell if I was entering or leaving the danger site. Then I saw black tape, which indicated a different kind of hazard, followed by another area of blue caution tape, indicating more hazard ahead. Trying to comply with these rules made my head hurt.

As we worked on jobs throughout the years, many stories were shared by my fellow workers. Some years ago, on a job at a state-run psychiatric hospital, a worker who witnessed a communication mishap shared the following story: A large dump truck was backing into position to dump a load of dirt into a ravine that the workers were filling in. As the driver was backing up, a co-worker was in place to assist him to back into position. The helper indicated for the driver to keep backing up. The truck eventually backed up so far that it fell over the side of the ravine and caused great damage to the truck. As the driver very angrily made his way back up the top of the ravine, he looked for the helper that had backed him up.

He could not find his helper, so he went in search of the little man in the hard hat, only to discover that there was no helper. It turned out the little man in the hardhat was a patient in the hospital just out for a stroll with his favorite hat on. The driver of the very expensive truck did not find much humor in the event. Fortunately, no one was hurt.

Over the years, using a motorized lift was a big part of our job installing overhead work. One job would not allow any step ladders on the job. All elevated work had to be done using a lift. Going up was not the problem. Getting the lift in place to go up was the problem. Imagine trying to maneuver a lift through narrow doorways to complete the task at hand.

It was often challenging to move a lift into place to perform our work. We were installing duct very high in the air. We had placed an elbow on top of the lift to move it into place. No one on the lift could stand up as the elbow took up all the room on the lift.

You could not see down but you could see the adjacent wall through the rails of the lift. There were four of us on the lift to make the connection in the air. As we moved up, I was following the mortar joints on the block wall, which normally would line up evenly straight up the wall. It appeared that we were starting to drift to one side, and I thought we were tipping over due to the big fitting on the top of the lift and my lack of being able to see over the side.

I started screaming to stop, but the man maneuvering our lift just kept going. He could see the ground and knew we were okay to keep going, but I could not see the ground and was fearing for my life. Everyone in the lift had a good laugh but me. Finally, when I realized that all was okay, I calmed down and went to work. Just another day on the job.

These were examples of things that would never happen on the job today. Safety and responsible actions are certainly justifiable.

I was only remotely involved in one fatality on a job where I was working. In Kansas City, it quickly circulated around the job that a fitter had been killed right before lunch. He was working in a lift and had pushed the lift extension out to work. He then started to move down, and the lift became snagged and finally tipped over. He was killed in the fall inside of the lift.

As an act of honor, the entire job stopped, and everyone left for the balance of the day. Somehow word made it back to the home office in Dayton. I was called and questioned as to why the crew had left the job for the day. It was an awkward conversation.

As an apprentice, I had the unfortunate experience of working though another experience with death on the job. This was well before safety would be strictly enforced. Our field superintendent came to our job and asked me to move to another job with a foreman I did not trust. I don't know how I had the courage to tell him I did not want to work with this foreman. Or how I could ask him to change my assignment, but I did.

Several days later, on my way to work, I heard on the morning news that there had been a fatality on a job the day before. It was the co-worker who had taken my place at the job I didn't want. I was stunned.

I traveled on to work and was told by my foreman the

superintendent was on his way to our job. He arrived and told us that we needed to be on the lookout for the OSHA rep as he felt certain that the rep would visit our job due to the circumstances of a fatality in our company. My foreman informed our superintendent that we would wear our safety glasses but not our hard hats. That showed how much our men did not comply with safety protocols, even at a time like that. They thought safety was something that could be tolerated but not emphasized.

The emphasis was not on working safely. The most important concern was to get the job done quickly. That would certainly change in the future. We would become one of the safest industries.

Activities that were common before the advent of OSHA were things like riding the ball of the crane to the ground at lunch and quitting time. It saved a long trip of climbing down to the ground. The operator would take as many as could fit their feet on the ball at once. It was a rush to do so. You could see a great distance as the operator playfully took you to a great height before dropping you to the ground.

As you might guess, there were injuries sustained as workers were either dropped or could not hold on to the cable. We would also wrap our legs around the steel columns and slide down to the ground. This also resulted in many accidents on the way to lunch or home.

Some workers found complying with safety regulations to be less than manly. No macho worker would want that.

Another occasion involved working on the demolition of a shot blast booth used to clean up foundry castings. We would next install a new booth. During the demolition phase, we were moving old pieces of the booth to the scrap yard and making good time moving the parts. I was in a heated conversation with the business agent of the operator's union as he was asking us to hire an operator to run the fork truck. There was not enough work to justify hiring a full-time operator to do part-time work. We could perform that task with our own crew.

As I was discussing this situation with the agent, I noticed our apprentice calmly standing by, waiting to talk with me. In a break in the conversation, I asked him what he needed. He informed me that there was a fire in the scrap yard. He wanted to know if it was okay to call for the fire department.

I broke off my conversation to ask if they had attempted to put the fire out. I was informed that they had emptied many fire extinguishers on the fire and were trying to use a 4" fire hose, all to no avail. They did not want to call the fire trucks until I said it was okay. I *ran* outside to see what was happening. As I left the building, I could see flames jumping up to about thirty feet and yelling for someone to call the fire department. This was prior to cell phones so it was not as easy as today to call for immediate assistance. The fire crew arrived with many fire trucks and firemen to put out the fire. Nothing was damaged as the fire had been in the scrap yard.

I informed all of those on site that it was okay to call the fire department whenever anyone thought it a good idea to do so. I also told our apprentice that it was okay to interrupt a discussion during an emergency.

Early in my career, we worked on dust collectors that removed dust created during the process of making brake shoes. This, too, was a product of Dayton. I had the privilege of meeting one of the engineers that worked on the original brake systems for automobiles. At the time, the process included shoes made of asbestos. As the material was cut and formed, it produced dust that was one hundred percent asbestos. It moved into a cyclone-style collector much like a huge shop vac. The bottom often bridged over and did not allow the material to be deposited into the screw conveyor to remove it from the plant.

The best way to address this issue was to crawl into the collector and shove a two by four into the bottom of the collector through the stoppage. No masks or any type of breathing apparatus was used. If asbestos is really that dangerous, then I should already be *dead*. It was the job of the apprentice to enter the collector. And I did.

As an apprentice, I did many risky things. On another occasion, I was to use an access door to enter round ductwork and crawl about twenty feet through the duct and into an industrial fan to replace its worn blades. I carried the new replacement paddle-looking blades into the fan to weld them into place. I entered the fan and started to work, welding in a very confined space.

As I was working, all I could think about was if I had locked out the electric supply to the fan from service. I could not remember with certainty. It was customary for the worker to place his own lock on the system lockout and tag it with his name, so others knew who was working on the piece of equipment.

I became quite concerned and moved to quickly remove myself from the fan and check to make sure that it was safe to continue working. I secured the lock and finished the job. No more fear of the fan starting and chewing me up. Stress is a part of the job.

An opportunity presented itself to me as an estimator and field supervisor. The project was the replacement of an industrial oven at a plant that made tractor trailer trucks. The job was located directly over the final assembly line. It was a summer job and was planned to be completed over the July fourth shutdown of the assembly plant. The project was progressing nicely toward completion. We were working a double shift to complete the job as quickly as possible during the short time frame with which to complete the task.

It was summer and it was hot. It was a time of pop-up showers quickly soaking an area. Just such a rainstorm happened, targeting our location. We could see it coming over the horizon and certainly could not stop the storm. It poured down rain. The rain collected on the roof and formed a river heading for the only opening on the roof: our roof opening into the oven.

As the water found its way into the plant, it also found its way into eight previously beautiful new trucks that were on the final assembly line. All were damaged beyond repair to the cost of approximately $90,000 each.

I received a brief but angry call from the engineer. As the field superintendent, I went to meet with the inexperienced engineer and reminded him of our prior conversation: my asking him to move the trucks before we started work. At the time, he said that they could not be moved so they were not.

The next trip was to the plant manager's office, who was not at all happy either. First, he straight out asked the engineer why this had occurred. The engineer did not have a satisfactory answer. The plant manager then turned to question me about what had happened. I then mentioned that pop-up showers "pop up" and I had requested the trucks be moved prior to our crews starting work. Then the longest thirty seconds took place as I waited to hear what the engineer would say. His response was slow but true. He admitted that he did not move the trucks as requested. Shortly after the job was completed, the engineer was fired by the company. I felt horrible, but nevertheless, our job was complete on time and under budget.

As I moved through my experiences in the industry, I was involved in a variety of locations and types of work. No two days were the same. We were once contracted to exhaust many chemical and experimental hoods in a college setting. I worked in several colleges over my career. Not every job was dirty and filthy. This job involved running the exhaust duct up through the roof, connecting all the outlets together on the roof, and exhausting them through a roof-mounted fan. This required a great deal of work to be performed on the roof.

As we were working on the roof, we could see for a long distance throughout the college town, and it was an idyllic setting. As we worked away, we could also see a trench being dug in the yard below for new underground pipe. The trencher was working away with the "chunk-chunk-chunk" sound of dirt being excavated.

Suddenly, the sound changed to a "clunk-clunk-clunk." The operator had hit something in the dirt. As we made our way to the edge of the roof, we could see a worker running across the yard and the hole in the ground spewing a vapor and much dirt flying in the air. We *immediately* smelled the strong odor of natural gas.

We had an excellent view of the disaster below. The next sound we heard was sirens from the fire trucks. They rolled up and looked in the hole. They saw no fire, so they left.

Soon thereafter, the maintenance men showed up to repair the pipe with a backhoe. They feverishly dug down to the pipe and were attempting to clamp off the leak. This was *after* they failed to close the main valve at the street.

As they tried to finish the excavation, the backhoe hit the pipe and poked a new hole in it. Now more digging was required. We had a great view of the events. Finally, the clamps were installed, and the gas was shut off. And no fire.

We went back to work for what we thought might be a great finish to a big problem. It was a little over a half hour before a representative of the college came up on the roof and informed us that we had to leave the building as the gas leak was not yet taken care of.

We reluctantly gathered our tools to stop work for the day leaving several hours early so the building could be cleared of the gas leak. Another day in the life of a construction worker.

The next job was in a skyscraper. This was for a bank being built in Dayton for their headquarters. Normally we would never have the opportunity for the local police to appear on our jobs--until this job.

They were *everywhere* and no one could figure out why they were present. Finally it occurred to us that we were building a bank, and on that day they were moving the safe deposit boxes into the new vault. No one had shared that information with us, and all at once we were flooded on the job with a strong police presence. It was the only time I can remember being prevented access to our usual work area due to the presence of the police force.

One tradition of iron workers needs to be shared here; it is called the rain circle. Iron workers were famous for drawing a circle on the roof of a project and waiting for five rain drops to hit inside the circle. It was then that they went home in the interest of safety.

Some jobs called for a very small circle to be drawn. It took a lot of rain to finally call off the day. Other projects saw a circle that was many feet in diameter. It took very little rain to call off the day.

A great deal of fun was had as we waited for the rain circle action. My brother, who worked as an untrained iron worker for several years through college, recalled the ironworker chant, "Lordy, lordy, if it's in your power, how about a quick knockoff shower."

Working with trucking companies for deliveries was often a nightmare. Working on a downtown job was also challenging. I received a call from a trucking company stating that our project had a unit that was required for our downtown job that was to be delivered the following day. We normally requested and received a 24-hour notice of delivery.

Notice was given and it was stated that the truck and unit were at the plant in Dallas, Texas. I questioned the ability of the company to deliver within the time allotted but was told that they would be there.

Before noon the next day, the call came that the truck was in front of the building. I made my way to the street and, sure enough, the team driven load was on the street. The truckers climbed out of the truck and were asking how long until we unloaded the equipment. We moved quickly to unload.

Not waiting very patiently, the team of two drivers ran laps around the truck. I don't know what they were on, but speed was of the essence that day. We unloaded the truck and off they went. Quickly.

Drugs unfortunately are everywhere, and construction is not immune to their presence. One of the shops where I was asked to serve as shop foreman had a reputation as a very tolerant place to work. As I was moving from the office, walking back into the shop as a new foreman, I noticed a line of crack cocaine laid out on the metal as two workers were welding a large piece of pipe.

I stopped them and indicated my displeasure with the activity. I told them that was their one and only chance. I never saw it in the shop again. New sheriff in town--new rules.

On a job that I was helping to install, I noticed the ironworkers go through the process of being drug-tested. This was many years ago when drug testing had just started on the jobsites. At that time, it took about a week to get the results, and the iron worker crew completely changed every week. One worker came out of the test trailer, threw up his hands and said that they were testing for five drugs, and he was proud to say he had them all!

Times have changed for the better and now the job site is much safer and drug-free. Many trades are required to pass random drug screenings and most larger jobs require a pre-employment drug test to access the job. The accident rates are lower on most jobs. It is a safer place to work.

This contrasts with a jobsite that at one time included alcohol as a way of doing business. As an apprentice, I can recall a time when we were working an aggressive job at a local auto plant that required working 16 to 18 hours straight with little regard for safety.

We would start work at 7:30 in the morning and work until morning break at 9:00. Break time included a trip to the bar for some down time. All the workers would down a pitcher of beer in the ten-minute break and return to work. We would work until 11:30, stop for lunch, and a trip to the bar. Food and another pitcher of beer.

Then we would return to work, until the 2:00 break. Another trip to the bar. Another pitcher of beer. Work continued until 4:00, our normal time to conclude work for the day. Another break, another pitcher.

We then worked until 6:00, when it was time for dinner break. Another pitcher and some food. Return to work with another break at 8:00 and more beer. The work continued with a break at 10:00 and more beer. Each time a pitcher of beer for the journeymen on the job. I could not begin to keep up and did not try.

The next break was scheduled for midnight, and we were on time for that break as well. At this break, the bar informed our workers that they closed at 2:00 AM. So we returned at 1:30 for our last break.

On this return to the jobsite, it became difficult to continue working as the journeymen were starting to have difficulty in proceeding. I still remember our crew walking arm in arm through the guardhouse, holding each other up as we returned to work.

We continued working until almost 6:00 in the morning as we finally completed our work in the plant. No more breaks as it became difficult to stop and start again when you were that exhausted (and drunk.) We completed many projects that required us to work the hours that are not safe to proceed; however, we somehow completed the work to the satisfaction of the engineer and customer.

Things have certainly changed on the job site. This kind of work ethic is not tolerated now and should never have been accepted, but it was at that time. It is an unexplained miracle that we all survived.

Many injuries were expected and accepted in the early years. As time has progressed, the building trades have become a very safe place to earn a living.

Chapter Six

Change Orders

Change orders normally result in a change in the contract amount for the work to be completed. At one point, change orders were a very lucrative part of a job. This is no longer the case on most projects. I remember receiving a contract on a project for $4000 and receiving $84,000 in change orders. It was customary to accept a job for a fixed amount and then add change orders for a large sum of money.

Although change orders used to be a major part of the work, that has changed over time. On most projects the customer now has limited the percentage of change order work allowed on each job. So, the $4000 project would now be limited to only $800 in change orders. The limit is now set at a predetermined 20% of the contract amount or whatever amount is determined by the customer.

Customers are now asked to make a list of the projects they would like to include as extras. As a contractor's representative, I can attest to the flurry of activity for many change orders at the start of a job as money is still available. As a contractor that is present for the end of the job, when money has dried up, I have seen change orders that are very difficult to ask for and receive.

Typically, a change order starts with an RFI, which is a *Request For Information*. It is a written request for clarification or a solution to an unclear description of the specifications or drawings. The contractor desires a change that results in additional work and a change to the dollar value or length of time for the contract. The customer, however, desires a clarification of the existing contract with no further costs incurred. It is important that the terms of the contract are clearly written to fully understand when a change is being made or added and what additional costs might be incurred because of the change to the contract.

There are two important elements involved in the possible change: time and money. A contract extension may be as important as the change in the contract.

Originally the contract is based on what work is to be accomplished, how much it will cost, and how long it will take.

For SMACNA, I taught a class nationally on *Change Orders*. One of the change order cases that was worked on in the class involved a contractor who attended the class. He patiently waited until the class was over and approached me to discuss his situation. The contractor was trying to finish a job, hoping that the conclusion of the job would incentivize the customer to generously offer the desired change order. He was hoping that the situation would be settled in the contractor's favor with additional payment for the work.

This contractor was stretched as far as he could be stretched. He had started a sizeable job that expanded into a larger project. The project generated over 200 RFIs, asking for change orders for the 200 requests. He received no response to any of the requests.

At that point, the bank entered the picture as he had borrowed money to keep his business afloat. Most small contractors are required to guarantee their business operation with their home, auto or other collateral on the line. The customer was stalling, knowing that the contractor would be in serious financial difficulty, as cashflow would be an immediate concern.

This was a classic example of a contractor trusting an untrustworthy customer. The bank continued to press for a swift conclusion. The contractor came to the training in hopes of saving his company. This situation occurs with contractors both large and small. I can recall traveling a great distance to pick up a million-dollar final payment and bring it back to my contractor that turned out to be the only way to cover payroll that week for a large contractor.

I suggested someone skilled in working with this small contractor to assist him through this problem. Fortunately, the problem had a successful resolution. The contractor worked with a former contractor skilled in working with banks. He worked with the bank and the customer to reach an acceptable resolution.

Today change orders remain a way of life for contractors; they continue to be a challenge to reaching an acceptable conclusion for every job. The two most difficult parts of a job are to start a job and completely finish a job.

Chapter Seven

The Way We Estimate

The world of estimating can be a tricky one. The estimator works very hard to include everything in a quote without overlooking the necessary details, all the while trying to be the lowest bidder. A guaranteed way to be the lowest bidder is to forget or overlook something. This can and does happen easily.

We estimated a job at a federal installation; we worked extremely hard to make sure we included all the items necessary to complete the job. And we had. I led the team through the process and felt good about the number we arrived at for the bid opening. It is standard in the construction industry for all larger projects to be reviewed by other estimators to make sure nothing is missed or overlooked.

The first bid that was opened by the contracting officer was from a good competitor of ours. It was thirty percent higher than our number. I thought that they were busy and did not want this job. The next number the contract officer read was even higher than the first one. Now I was worried, concerned, and a little afraid.

I was trying to tell myself that neither company wanted the work. How could I be that wrong? There were four bids to be opened for this job. Each bid is time-stamped as it is received. They were opened according to the time-stamp. I knew ours would be opened last.

The third bid opened was two percent higher than our bid, a perfect place to be *IF* the third bid was correct. Or were the numbers on bids one and two, correct?

When our number was read, our quote was the apparent low bid. It was just over the number of the next lowest bid. My first thought was one of relief that we had the lowest bid with the correct numbers. My second immediate thought was all about wondering if the other low bid may have also had incorrect numbers.

Off to the office I went with the news that we had a nice project to start very soon. I arrived at the office to learn that we needed to secure our bid with a bid bond guarantee. Then I was told that the company was out of bonding ability because we had previously used our available bonding on other jobs.

Weeks of work went down the drain as the job was awarded to the second-place bidder. Bidding and estimating work are sometimes very frustrating.

The other way to secure work through the bid process is to forget to include something. We started a rather large project in a major town just to our south. I was assigned to be the project manager for the sheet metal work. We had the mechanical contract on the job, as well. That meant we were to perform the plumbing and mechanical piping for the project as well as the sheet metal for the air conditioning and heating.

The project was an 1880s hotel being transformed into a state-of-the-art downtown hotel. The building was seven stories high. The basement housed the kitchen for the first-floor restaurant, and the second floor was for meeting rooms and a ballroom. The top floor was a mechanical room for the hot water boilers and other mechanical equipment. The remaining three floors were identical guest rooms.

The bid was awarded for the project to our company. It has been completed by our bid team of professionals with many years of experience. It was two percent lower than the closest bid. Our team felt comfortable with the job. I was selected to perform the role of the project manager for the sheet metal portion. A separate project manager was used for the mechanical portion. As things began to move forward, it was evident that something was wrong. I was pressured constantly to perform better than we had been doing. I was tracking the project, and the actual pounds per hour were very close to the normal bid numbers. All looked good to me.

But I was constantly driven to do better. Finally, I was let in on the secret: We had bid on only one floor of the three floors of the repeated guest rooms. The bid team had missed two floors of the job. *Both* sheet metal *and* plumbing!

Now the trick was to recover from that major oversight. It is easy to secure a job but very difficult to make money when you make an error like that one. I reviewed all the bid documents but had not caught the error on my first look either.

Another job that I was heavily involved with was a "clean

room" that required a great deal of air filtration and movement. I looked over the scope of work and decided to assemble sections of ductwork finished with all the usual field-installed and assembled parts and pieces.

We delivered the assembled pieces to the job at much longer lengths than are normally sent to the site. We moved the assembled lengths into position and lifted them into place with minimal work done in the field. Normally the opportunity to move the assembled pieces into place is not an option.

This reduced our field time significantly, but we ran up our time in the shop. So, I was called into the office to explain why our shop time was so high. No one had bothered to look at the reduced time in the field. The job was successful as the total time was well below the budgeted time allotted.

This was an example of approaching the same job differently through the substitution of more shop time as opposed to less field time. This is not possible for many jobs, however, as access to the site is often restricted and very congested. The average number of moves on the job for a piece to be installed into its final resting place is *seven* times! If you install it before it moves seven times, you are making money.

This same situation occurred on a job for a long, enclosed chamber we were building for a paint spray booth at an auto plant. There are many ways to get to the end of a job. This job was overseen by an engineer that had witnessed a crew assembling the booth from the bottom up at a previous job.

We, however, had chosen to build the top plenum first, raise the top up, and then stand the walls up in place. This method sounded better to our team. We were taking that approach, and the engineer had never seen this method. I was hoping this would work better.

As it turned out, we appeared to be a week off schedule according to the engineer, when we were partially complete. The engineer called me into the office to inform me of this information. The engineer and contractor really wanted to stay on schedule. As a result of us taking a different route with this task, we were able to complete the installation a week early. Unfortunately, no atta-boy was offered. We completed the installation.

There are many options for manufacturing and assembling duct in various forms including transition fittings, elbows, and offsets. Some shop processes require straight duct to be fabricated in four-foot lengths. Some allow ductwork to be fabricated in five-foot lengths. Neither is better or worse, just different. It all depends on what equipment you have as your shop's standard. There are less pieces to install as the duct length grows. All need to be compliant with SMACNA (Sheet Metal and Air Conditioning Contractors National Association) standards. The industry standard *Duct Construction Manual* can be followed and satisfied in many ways. It is an extensive manual of options, full of many different methods of constructing duct work.

The manual was produced with the idea that industry needed to train individuals to estimate work with a basis of knowledge to learn what drives the computer-generated numbers that the systems use today.

Through my SMACNA affiliation, I had the opportunity to work with a talented, skilled and experienced group of contractors to develop the *Basic Sheet Metal Estimating* manual. The manual was published by SMACNA and continues to be used today. I had the pleasure and honor to write the manual while developing a course on estimating through the SMACNA chapters across the United States and Canada.

The group that formed the committee to assist in writing the manual contained contractors from different areas of the country. We met several times at various locations. The committee was extremely helpful in offering topics, assisting with writing, and editing the material.

Younger workers are relied on and comfortable with the numbers generated by computers but did not know what the numbers were based on. The course was designed to dive into the source of those numbers, which are taken from real-world contractors' numbers.

Contractors used different ways of getting to the end result. Some relied more heavily on extra shop time to reduce the field time on a project. Others were satisfied with using more field time to make up for the lack of shop-inspired effort.

One exercise developed by the committee to be completed during the training worked quite well. We created an exercise that could not be accomplished with the information provided. The goal of the exercise was to show how easy it is to miss key items in an estimate.

The correct answer was to not offer to bid a job that could not be accomplished. During one of the sessions, I was working with contractors, not their workers, as I normally would do. The contractors were divided into groups of "estimators" and told to work through the exercise.

One group, even after being shown what they had missed, insisted that they would still take the job and somehow make money on the project. The example worked to perfection. It showed how hard the world of contracting can be.

Chapter Eight

Phase Two: Education

Phase One of my stories focused on sharing my experience through jobs and places that I have worked and the many challenging days I encountered through my career.

Phase Two moves into another chapter of my career. This phase is where I began using my experience to teach about my industry. About twenty years into my career, I attended a foreman training class and told the local SMACNA chapter executive from my area that the training was well done but thought I could do better. He challenged me to do so: I had no intention of doing that. My job as shop foreman was good and there was no desire to leave that position.

As it shortly turned out, I would not be in that position much longer. The field superintendent felt I was after his position, even though I was not. He waited until the division manager was on vacation and then had me terminated. I thought back to my desire to work with contractors/owners to improve their operations.

As I visited with owners, I kept hearing a similar statement;

they desired better foreman and leadership training. I went back to the contractors' chapter executive and offered to develop a foreman training session. He quickly agreed; after developing the session, I taught what was possibly my most difficult class: my own local union. The class was full of individuals that all knew me and had either worked for me or I had worked for them. A real tough group.

I survived that group and was then, unbeknownst to me, recommended teaching the session at the national level. I agreed to the challenge, not knowing what that might look like. Soon after, I had a California chapter contact me regarding the possibility of traveling out west to teach the foreman training class.

I agreed and went to work on writing the curriculum for what I felt the leader on the job needed. I was soon considered an expert as I was traveling a hundred miles or more to present the training class.

This class went well, and I was selected to offer more trainings the following year. Off to the races I went. With a half dozen dates scheduled by chapters across the country, I continued to improve by updating more and better material with suggestions from those in attendance and on the jobs that I was continuing to run on a daily basis.

This eventually morphed into six different leadership classes, all taught through SMACNA. Eventually, I would work with over sixty of the over one hundred chapters of SMACNA. Presentations at conventions and mid-year meetings would follow.

I ended up traveling over 950,000 miles in roughly twenty years working with many skilled people in various locations. The classes I taught eventually took me out of the country. This led to working with many trades and further led to presenting leadership skills to foremen, general foremen, and construction managers in many parts of the country.

In British Columbia, one of the attendees traveled over a thousand miles for the session. Talk about feeling pressure!

A trip out of the country helped me realize that jobsite problems were the same no matter where you went to work. Crossing a border did not affect the course of the work. I had feared that crossing over into Canada would be problematic as Canada was on the decimal system and not feet and inches. But I quickly discovered that most of the Canadian sheet metal workers did so much work in the United States that they too, were on the feet and inches system.

I also made many trips to Alaska to train workers. On a trip to Alaska, I watched the start of the Iditarod dog sled race on a day off from training. And yes, the problems on the job were similar, just colder.

While in Florida working with the tinners on October 29, 1998, I watched a space shuttle launch, even if it was miles away. Many in the training had worked at the launch facility yet had never seen a launch. We stopped our training and moved to the roof of the hotel training site to witness this amazing event. The launch had special meaning to me because John Glenn was the first American to orbit the earth, oldest man in space, and an Ohio boy was going into space again. I, too, am an Ohio boy.

Another interesting trip that presented itself to me occurred in

2002 as I was requested to travel to New York City area for a training on the one-year anniversary of 9/11. This was a return trip to New York as I was asked shortly after 9/11 to train just as air travel was opening up again after the terrorist attack on September 11, 2001.

On the day the planes struck the towers, I was working in the Richmond, Indiana area at a small college. I was moving around the campus working on different buildings. It was a stunningly clear day with bright blue skies. As I returned to our main project, I noticed a television set that had been set up in the space outside of our work area. The national news was on with a developing story. One of the students watching the news was quite troubled as she realized what was happening knowing that her father worked in the north tower. I never heard any more about her story and to this day wonder what the outcome had been or how the events changed her life.

As I worked with the group just north of the city, I realized their wounds were still very raw. Many, if not most, had personal connections to the disaster. Even though that day was very much on the mind of everyone in the session, I had no idea how personal it was in the New York City area. I developed a new understanding of how this event changed our country.

I had the opportunity to visit Ground Zero after the training on the first anniversary of 9/11 and had just missed the president on his visit to the memorial. This experience was due to the opportunity given to me through my association with the contractors' association and their local chapters.

It was a day that I will never forget. Seeing New York City after that unforgettable day was overwhelming. I flew into the area to empty airports and closed hotels as the recovery was taking place. It was an experience that is difficult to forget as my cab traveled through vacant streets with many closed hotels and businesses.

Although I first entered foreman training for the sheet metal industry, I moved into working with almost all the crafts and trades, including plumbers, fitters, electricians, operators, carpenters and ironworkers. As I shared laughingly with the roofers, I even tried to train them.

There are quality people in all the trades and crafts. Working to fabricate and install complex building systems requires quality people in all the trades working together. Coordination of the trades is essential to each project.

I was on my way to becoming the only union card-carrying member training other members in the industry. This gave me a leg up on other trainers who were teaching theory only as compared to my actual hands-on experience. I returned to many areas to train year after year.

My teaching took me to most major cities in the country. I made several trips to Alaska and repeat trips to several regions of the country. I developed many lasting friendships over the years of working with many areas. At times it was challenging. I continued to work in the industry and was lucky to have employers and crew members that allowed me to work, teach, and develop new sessions based on my work on the job.

In addition, working with different crafts and trades allowed me to develop the material for workers across the trades. Once again, people are people no matter what trade or area in which the work is completed.

The only trade specific material I taught was the estimating session for the sheet metal trade. The sheet metal industry is so diverse, the options are endless for different methods of producing quality products that all meet industry standards.

Chapter Nine

Why won't you listen to me? The need for change

The pecking order on a job usually includes workers who physically perform the tasks required, a foreman to oversee several workers, and a general foreman who manages the foreman and the workers. The general foreman usually reports to a project manager that may supervise multiple trades and may or may not be present on the job daily.

Several changes that have taken place over time can be frustrating. One job included a project manager that replaced the general contractor. He would not allow the general foreman from the individual trades to communicate with each other. Each contractor supplied a project manager that attended weekly meetings. The project manager was to report to the foreman of each trade.

The construction manager sometimes oversees unskilled, untrained younger workers. This can lead to a manager who has difficulty managing those who are reporting to him or her. It is often an engineer without on-the-job skills and often is a college-educated individual trying their best to manage workers, with years of training, that are quite capable of working without outside supervision.

A classic example of such a project was in a local paper mill. The individual foremen were not invited into the progress meetings. Keeping the trades separated was extremely difficult as we were unable to coordinate the job as we would normally operate. We were finally brought together for the first time when the customer chose to split the work on the electrical phase of the project between two contractors, one that was union and one that was non-union. Friction started soon after.

The man-lifts that were on the job were being vandalized; someone was cutting through the service cable that operated the lifts. This was obviously *very* dangerous, so the general foremen were called together to address the situation. I immediately brought up that we were not permitted to talk to each other to coordinate the project. When a matter of danger occurred, then we were to oversee our people. I stated that it was very frustrating not to be valued until a major occurrence took place.

After the meeting, we were allowed control of our workers to complete the project. The project was completed on time. With no additional safety problems.

The project manager on the job infuriated the workers through his callous disregard for them *or* their well-being. We were working 24 hours a day on the job. Two twelve-hour shifts overlapped so the project moved forward quickly. It was summer, and the job was very hot during the July heat wave.

One day, a worker went down with a medical emergency, likely a heat stroke, due to the extreme heat. When the squad was called to the site for the emergency, the EMT's arrived to perform first aid. The project manager stopped the first responders and would not allow them to perform first aid until they had put on hard hats and safety glasses. All this while the worker lay on the ground waiting for assistance. This prompted a meeting of the workers to complain to the owner about the callous way of working through an emergency.

The quality of the worker increased throughout the years I spent in the industry. At times, though, the quality of the *intermediate* worker with no trade experience that has entered the world of construction has not kept pace. There were times that I was directly supervised by an engineer without any mechanical trade experience, which was frustrating because the individual was making decisions that directly impacted my work and the work of those under my control.

The construction industry does a great job training our people to accomplish work despite working under managers less skilled and without on-the-job experience in construction.

Chapter Ten

Where Do We Go from Here?

The discussion repeated quite often for many years has been the anticipated need for more construction workers. The need is here now. During a recent local union meeting, the work in the local area was being discussed. On super jobs (jobs over one billion dollars) the amount that is now under contract in southwest Ohio alone is over 135 billion dollars. Work is being delayed in some cases due to the lack of workers.

Over time, the building trades have been manned on a just-in-time case. Looking to the future was not part of planning ahead. No future thinking was devoted to the cause. This is not the first time this has occurred in the industry. Previously I discussed the issues in southwest Ohio dealing with untrained workers on the job in the late 60's. The large Armco project moved forward and many cab drivers, bartenders, and other blue collar workers manned jobs during that time. It took many years of hard work to cycle enough workers through the required apprenticeship programs that are in place to train workers. Entering the construction industry now, for a lifelong career, is an excellent choice.

As I trained workers in the leadership training classes, I would ask three questions of the foreman level attendee. First, how did they enter the trade? Who brought them in and how did they find out about the industry? Second, I asked the leaders how the trade had been to them overall. All individuals thought the sheet metal trade had been good to them and it was a good career choice. I normally received strong positive responses to the question with many indicating that the trade had been an excellent career choice. The third and final question was asking for individuals who had teenage sons or daughters. I identified over three thousand workers during my twenty years of foreman training around the country, as having high-school aged children. Only five out of three thousand indicated that they wanted their children to continue the tradition of employment in the trade. This is *after* identifying almost ninety percent of the current industry as entering the field through a family-member recommendation. Most entered because they had a dad, uncle or friend who worked the trade.

I then shared with this group that not all children should enter college, and the numbers supporting this view were shared. The cost of a college education is approximately $25,000 per year. It normally takes *five* years to get a four-year degree. Your average debt is somewhere north of *$200,000,* often over *$300,000.* You will then qualify for a job paying on average $45,000 with a four year degree. Statistics indicate that you will have a 46% chance of securing employment in the field of study that you were trained in. You have a 29% chance of securing a career in a different field of study. And, you have a 16% chance of being *unemployed* with the degree *and* the debt.

This data should be compared to an apprenticeship program where you are paid to go to school, and you have a career in which you will *earn*, conservatively, $200,000 during your four years of apprenticeship. And you have 100% guarantee of a career for which you were trained when you complete the program. *And* you are trained with zero debt.

	COLLEGE	**APPRENTICESHIP**
DEBT	*$200,000 - $300,000*	*$O (NO DEBT)*
EARNINGS POTENTIAL *DURING 4 YR PROGRAM*	*?? (DEPENDS ON CO-OP OR PT WORK)*	*$200,000*
AVG. STARTING SALARY *AFTER FINISHING*	*$45,000*	*$80,000 (Over $100,000 with OT)*

That is almost a half-million-dollar swing for parents. When this information was shared with the group, it was amazing how many reconsidered their beliefs about the building trades. Nearly 100% agreed that the trades had been good to them over their lifetime.

High school career centers are now working toward offering students an excellent option to enter the trades. The construction industry still fights the stereotype of employing the hard drinking, hardworking and less intelligent individuals. This is no longer the case for the contractors fighting for quality workers.

The opportunities are limitless in the construction industry today. With computer and chip manufacturing returning to the United States, it will be necessary for the industry to work on many clean room installations. These installations are built to provide an atmosphere for the electronics to be built in a large facility that monitors the air quality and are required to constantly filter the air to a higher-than-normal standard.

There was a clean room close to Dayton to originally produce picture tubes and computer monitors. The job employed many sheet metal workers to build that operation. That plant closed and the work left the country. Fortunately, the computer chip industry is returning to the United States and is making itself present in Ohio as a new chip factory is currently being built in the Columbus area. It is an over twenty-billion-dollar job, which will impact Ohio construction for years to come.

As you enter these facilities for electronic work, it is necessary to go through air locks to maintain a very clean atmosphere, requiring many trained construction workers as the projects move forward.

A massive amount of construction work is required as the country moves through the need for infrastructure improvements. Bridges, roads, and utility work are in need of trained workers as we continue to improve our country as we move into the future.

Epilogue

The construction industry is a great option for the younger generation entering the workforce. The industry has a bright future full of many areas of great need. It is also a very economical entrance into a career.

I spent many years in the construction industry taking me to places and opportunities new to me. Construction work varies day to day and is an excellent opportunity to build the economy. This industry, with improved safety measures, is a challenging way to earn a comfortable living.

Construction offers a different look at your world. You are literally building the towns where you live and work. As I pass by buildings where I worked, I often remark: "**I built that**" and often feel like I built it all by myself.

On *Share a Career Day* two of my sons came to work with me for a day. Both times provided interesting insight. My oldest son went to work with me on one of the tallest buildings in Dayton. This building was being constructed for a bank headquarters. My contractor was generous enough to allow my son to come to work that day. My son was a big boy and fit right in with the other workers on the job. At one point, my son was issued a hard hat and instructed to stop traffic on the street so a delivery truck could back into the truck dock.

He did his job and was shocked that the traffic stopped for a middle school kid. But stop they did. He was excited to be able to come to work with me that day and I was proud to show him my trade. He was not so brave when we went up to the roof to see Dayton's skyline. He wasn't alone as a few of the others on the job were also leery.

My third oldest son came to work in the winter; it was dark in the morning as we left for the jobsite. On that brisk morning, we arrived at the site and were met by a hearty coworker. My son was asked if he was a Hoover. When my son answered to the positive, the coworker informed him that he did not like Hoovers. It scared my son to death, but he was relieved when the coworker started laughing and by the end of the day both were fast friends.

Sheet metal is an occupation where you can easily see your part in the bigger picture: you see what you have accomplished by day's end. Many occupations do not provide this opportunity. If you push papers for a living, at the end of the day, all you see is the pile of papers.

Our youngest of four sons has entered the building trades as an electrician. He too has had the opportunity to work on some fascinating projects. He is currently working as a government contractor, wiring projects in a lab at a government research facility.

I make hand-made items, install the handmade items, and see systems that I have installed work through removal of dust, fumes, and metal shavings or wood chips. I also feel conditioned air moving about a building, helping to make the building a more comfortable place to work and live. I am pleased to be a part of the team that ensures that the building works as it was designed.

Construction allows the electrician to see the lights come on, the plumber to see the building operate with clean water and successful removal of waste, the carpenter to witness the walls support the structure, and other trades to see and feel their contribution to the project.

It really is a feeling of "**I built that!**"

The Rest of the Story

As a child, I grew up in the era during the daily radio news program with Paul Harvey. The host ended with a segment called, "*The rest of the story*." The program highlighted interesting facts in a story, followed by the ending: *the rest of the story*.

Here is my "the rest of the story." In June of 2024, I contacted the business agent at my union hall. I needed a favor. I wanted access to a sheet metal shop for a small project. He asked if I was interested in going back to work. My immediate answer was "no." I had been retired nine years and was not interested in returning to the workplace. Every year in January I receive a letter from the pension board informing me that if I worked in the industry, I would lose my pension. I had no desire to lose my pension income.

The business agent further informed me that the rules were changing that month and that I *could* return to work with my full pension and a full paycheck. At this point, I was mildly interested, but still said no. After further discussion I discovered, if interested, I would return as a supervisor and not used as a manual laborer. I asked where I needed to travel for this job. Earlier in my career, I traveled to work many times while driving over an hour to the site. I knew there were big jobs going unmanned an hour or more from my house, and I was not interested in driving that far.

I was happily informed that the job was at an auto parts plant being built in my hometown two miles from my home. In fifty-two years, I only worked in my hometown one day on a job. This made it very compelling to consider what appeared to be a great offer. I finally accepted the offer and became the first retiree in my local union to return to work to assist in the shortage of construction workers.

So here I am on the job for the first time in nine years. I was interested to learn if anything had changed and if so, what was new and different. I started out by going to orientation for the job, attending safety training for the construction manager, safety training for the contractor my contractor was sub-contracting with, and safety training for my contractor. A total of three back-to-back safety courses. Then I headed off to be drug tested.

That concluded my first day on the project. I certainly felt safety trained. Because I had passed my drug test, I was now allowed *on* the project.

My first day on the job site started in the morning with a group meeting of trades working on the site. It began with "stretch and flex" for all of those in attendance: touch your toes, bend over and the like. Then it was on to the morning briefing for the day's schedule. All activities were shared with everyone for each contractor responsible for their work. This included what truckloads of material might be received for the day.

We would then separate for a meeting with our own contractor where our safety officer would share the safety concerns for the day's activities. All of this before we started work. It was July, and it was hot. We were warned to make sure we stayed hydrated. This included popsicles being available for the workday. I was trying to imagine what the journeymen of days gone by would have thought about the availability of popsicles on the job.

I have now completed several months on the job. It is fascinating to interact with an out-of-town contractor who is employing many workers to man this immense project. It has been a great opportunity to work with new apprentices, newly organized workers in the industry, as well as with those who have some experience in the sheet metal trade. I have had the opportunity to work with probably the most skilled female operator I have ever encountered in my many years on the job. For that operator to be female speaks immensely to the changes on the jobsite over the years. This job has offered me the opportunities to meet and work with an impressive female apprentice who has a great work ethic. It is easy to teach the necessary job skills, but it is fulfilling to see the desire to succeed with a strong work ethic as we work toward completion of the project.

During my time on the job interacting with workers, many have shared just how good the construction industry is right now. A journeyman sheet metal worker shared that after earning more than $125,000 last year, he is taking some much-needed time away from the job. I also talked with a first-year electrical apprentice who entered the industry after working as a commercial painter for ten years, stated that they earned over $75,000 last year on first year apprentice wages. My grandson has entered the trades as an electrical lineman and will top out after four years as an apprentice at over fifty dollars an hour. His next-door neighbor entered the trades as an apprentice plumber/pipefitter and will earn well over what his mother is earning with a four-year degree in nursing and twenty years' experience. The opportunities to say, **"I built that**!" are unimaginable.

Working again as a supervisor further afforded me the chance to interact with the construction manager and other trades as I go about installing duct for a thirty-acre-under-roof building on a very aggressive schedule. Having built similar projects allows me to compare this very fast-track project to other projects. This is the quickest that I have followed a structure this size built to completion. It has been a clean, well-supervised, and safe project.

Working with other trades on the job offered me the time to talk with today's workers as we work to complete the building. Many traveled a great distance to work on this job.

This job allowed me to listen to workers during our morning group meetings present entirely new ideas to me and the rest of the crew. It was often stated that the construction manager on the project understood that the job required long and demanding hours. Building tie-ins from the existing building utilities would need to be done on holidays and weekends, causing an employee to work many hours on consecutive days. This prompted the need for holiday pay and was rewarding for the workers as holiday pay of double time became the reward for consecutive days worked.

As I experienced many times in the past, craftsmen would be expected to show up every day to work, with no exceptions. The paper mill project that I worked on, for example, expected everyone to work two overlapping twelve hour shifts so that construction was constant. If you did not show up for work, you would be fired.

With my current job, for my first time ever, in the morning meetings, it was stated that if anyone needed a day off, he or she should unquestionably take the day off. This was certainly a huge change in attitude towards time off.

This is a time in our history when the much talked about shortage of construction workers is right on top of us. Many jobs, like this project, are paying bounties just to show up for work. It has created a problem as many area jobs are offering these bounties due to the shortage of skilled workers. Several jobs are discussing the possibility of double time for *all* hours worked just to compete for the available tradesmen.

This has also caused jobs to be put on hold. The customers have realized the prices of the construction projects have skyrocketed, because the tradesmen are in such high demand.

Many workers are surprised to learn that the education to learn the building trades through an apprenticeship program is offered at no cost. As the cost of college education continues to soar without a guarantee of employment upon college graduation, apprenticeships are one of the best deals available.

As the United States' infrastructure continues to fail, more skilled trades will be required. Again, the possibilities are endless for the skilled trades.

We continue to need more craftsmen who can say,

"I built that!"

For Additional Information

The construction trades include the following individual skilled trades.

Carpenter-Constructs, installs, and repairs wooden structures

Electrician/Wireman-Tradesman specializing in electrical wiring of buildings. Also repairs existing electrical items

Electrician/Lineman-Installs electrical transmission lines and towers

Plumber/Pipe Fitter-Installs, repairs, and maintains piping, fittings and fixtures used in water supply, sanitation, and heating systems

Sheet Metal Worker-Fabricates and installs items made from sheet metal including ductwork, metal roofs and specialty items

HVAC Technician-Installs, maintains, services, and repairs heating, ventilating and air conditioning systems

Ironworker-Builds and installs the metal framework of structures like bridges, stadiums and buildings

Painter-Paints buildings, walls, ceilings and woodwork

Millwright-Installs, maintains, and repairs industrial machinery

Operator Engineer-Operates a variety of power machinery including forklifts, bulldozers, backhoes, cranes and other mechanized equipment

Cement Finisher-Places concrete in forms and finishes the concrete to a smooth finish.

Insulator-Places insulation on pipes, ductwork, and sheet metal to provide for material shielded from moisture and heat

Laborer-Provides manpower to areas that do not require skilled trades to perform the work

For more information on construction apprenticeships, below is a list of contacts. *YouTube* and *TikTok* also provide many good videos of the construction industry.

Carpenters *United Brotherhood of Carpenters and Joiners of America* *www.carpenters.org*

Electrician/Wireman *International Brotherhood of Electrical Workers* *www.ibew.org*

Electrician/Lineman/Power Line Workers *International Brotherhood of Electrical Workers* *www.ibew.org*

Plumbing/Pipe Fitting *United Association* *www.ua.org*

Sheet metal *International Association of Sheet Metal, Air, Rail and Transportation Workers* *www.smart-union.org*

Heating, Air Conditioning and Refrigeration *SMART union* *www.smart-union.org*

Ironworkers *International Association of Bridge, Structural, Ornamental and Reinforcing Ironworkers* *www.ironworkers.org*

Painters *International Union of Painters and Allied Trades* *www.iupat.org*

Roofers *United Union of Roofers, Waterproofers and Allied Workers* *www.unionroofers.com*

Millwrights *United Brotherhood of Carpenters/Millwrights* *www.ubcmillwrights.org*

Operators *International Union of Operating Engineers* *www.ua.org*

Cement Finisher *Operative Plasterers and Cement Masons* *www.opcmia.org*

Insulator-*International Association of Heat and Frost Insulators and Allied Workers* *www.insulators.org*

Laborer-*Laborers International Union* *www.liuna.org*

About the Cover

The Generations statue shown on the cover is located in Falls Church, Virginia, at the offices of the Sheet Metal Workers National Pension Fund. This eleven-foot-tall bronze statue depicts three generations of sheet metal workers, emphasizing the transfer of tools and knowledge from older family members to a young apprentice. The artist who created the statue is A. Joseph Kinkel, whose own father and grandfather worked in the sheet metal industry and served as models for the sculpture. Further details about the sculptor can be found on his website at www.ajkinkelsculptor.com.

About the Author

Roger Hoover is a member of the Sheet Metal, Air, and Rail Transit Workers International Association (SMART) Local Union 24. He attained journeyman status in 1976 upon completion of a comprehensive four-year apprenticeship program in Dayton, Ohio. Roger was first introduced to the trade during his senior year of high school through a visit from the Local Union to his drafting class. After attending Wright State University for one year, he determined that pursuing hands-on work in the trade aligned more closely with his interests.

He has dedicated fifty-two years of service to the local, holding positions such as field foreman, shop foreman, project manager, and general foreman on various industrial, commercial, and HVAC projects. Throughout his career, he has collaborated with business agents, owners, contractors, and a wide range of committed and highly skilled tradespeople.

Roger's practical experience and passion for his craft motivated him to share his expertise by facilitating training and leadership courses. Through the Sheet Metal and Air Conditioning Contractors National Association (SMACNA), he provided instruction to leaders across the sheet metal, electrical, roofing, plumbing and fitting industries. He also authored the Basic Sheet Metal Estimating Manual, a resource used within the sheet metal industry.

I Built That! and So Can You is a book that shares his experiences in skilled trades and discusses the importance of encouraging others to pursue careers in these fields.

Roger is well known in his community for his contributions as a volunteer. He has been a member of the Optimist Club of Brookville for over forty years. Roger is also an active longtime member of Community United Methodist Church. He dedicates his time as a scorekeeper at high school basketball games, assists with parking during football season, and actively supports his grandchildren by attending their sporting activities.

www.ingramcontent.com/pod-product-compliance
Lightning Source LLC
LaVergne TN
LVHW022055200926
848481LV00003B/927
* 9 7 9 8 8 9 3 9 0 0 7 1 2 *